MATHEMATICS
AND THE UNEXPECTED

MATHEMATICS
AND THE
UNEXPECTED

IVAR EKELAND

WITH A FOREWORD BY FELIX E. BROWDER

THE UNIVERSITY OF CHICAGO PRESS

Chicago and London

IVAR EKELAND is professor at the University of Paris–Dauphine, and has also taught at several North American universities.

The University of Chicago Press, Chicago 60637
The University of Chicago Press, Ltd., London
©1988 by Ivar Ekeland
All rights reserved. Published 1988
Printed in the United States of America

97 96 95 94 93 92 91 90 89 88   5432

A translation by the author of his *Le Calcul, l'Imprévu: Les figures du temps de Kepler à Thom.* ©Editions du Seuil, 1984.

LIBRARY OF CONGRESS CATALOGING-IN-PUBLICATION DATA

Ekeland, I. (Ivar), 1944-
    [Calcul, l'imprévu.   English]
    Mathematics and the unexpected / Ivar Ekeland ; with a foreword by Felix E. Browder.
        p.   cm.
    Translation of: Le calcul, l'imprévu.
    Bibliography: p.
    Includes index.
    ISBN 0-226-19989-4
    1. Catastrophes (Mathematics)   I. Title.
QA614.58.E3413 1988
514'.7—dc19   87-30230
                                                                 CIP

CONTENTS

FOREWORD

Ivar Ekeland's book *Mathematics and the Unexpected* is a translation by the author of his 1984 book in French, *Le Calcul, L'Imprévu* with the subtitle *Les Figures du temps de Kepler à Thom*. It is a popular exposition of a family of mathematical ideas and results about the mathematics of time (as the subtitle indicates) which go back to the work of the great French mathematician and mathematical physicist Henri Poincaré at the end of the nineteenth century and which have become one of the major themes of contemporary work in the physical sciences and engineering under such labels as *nonlinear science* and *order and chaos*. Ekeland's book in French was hailed as one of the outstanding works of scientific popularization of recent years and received the Jean Rostand Prize of 1984 for works in this genre, awarded by the Association of Science Writers of France.

Good popularizations of mathematics are exceedingly rare, rarer by far than in the physical or biological sciences. This was remarked some years ago by the well-known essayist and writer on physics Jeremy Bernstein, in a *New Yorker* review of a book of expository essays on mathematics, *Mathematics Today*. The situation has not changed fundamentally in the ensuing period. Although a number of books have appeared and received a reasonable amount of attention and popular reviews, one, *The Mathematical Experience*, by P. Davis and R. Hersh, had the important dual advantages of being attractive to the general reader and yet giving a reasonably accurate view of the mathematics presented. Of the others, some of the most readable either had massive errors or ground an ax with excessive zeal, or both. Although Ekeland's book is not intended to be a systematic overview of mathematics as a whole, it makes an exceedingly valuable contribution to the literature of mathematical popularization. It is very readable, as readers can discover for themselves, particularly in the way that it combines the genre of the scientific popularization with the somewhat different genre of the philosophical fable. It concerns a subject matter of great current importance in the development of science which promises to be even more important in the future, namely, how the simple mathematical laws which govern the basic physics of the universe give rise to the complex and even chaotic behavior of the actual world. Ekeland's book is accurate, and by any reasonable standard untendentious, although it deals in some of its parts with matters of controversy.

One should say a few words about the author. He is a professor of mathematics at the University of Paris–Dauphine, which is in effect the Business School of the University of Paris, and director of CEREMADE (Centre de recherche de mathématiques de la décision), which is the research group in what Americans would call operations research or management science. He has a distinguished reputation as a mathematician specializing in two principal areas of research of a relatively applied character, nonlinear optimization and the study of periodic solutions of Hamiltonian systems. These subjects are not what this book treats, and Ekeland is not following the all too familiar pattern of authors who write to praise the merits of their own work. To my knowledge, he has not done research on chaotic dynamics, but he is obviously able to expound the basic ideas of this subject and its intuitive implications far more expertly than the specialists and with commendable enthusiasm and good judgment.

His unusual personal background may well have played a major part in generating his unusual independence of mind. The son of a Norwegian father and a Corsican mother, with his father in the Norwegian diplomatic service, he spent a good part of his childhood in such exotic places as Vienna, Moscow, and Minneapolis as well as in Paris. His education culminated in the Ecole normale supérieure in Paris, but the English of

this book may well have elements of Minneapolis in its formation. His mother was a teacher of philosophy in a Catholic
spiritualist tradition, and it is only this combination of circumstances which could give rise to a French mathematical aca-
demic with a penchant for reading treatises on Neoplatonic
philosophy in Norwegian.

The two patron saints of the book are Poincaré and Bergson.
It was Poincaré, as Ekeland tells us in the early chapters, who
in his famous book on celestial mechanics showed that such
relatively simple systems as three bodies moving under the
control of gravity as given by Newton's gravitational law could
generate behavior of incalculable complexity and even disorder.
The mathematics of long-time or *asymptotic* behavior only de-
scribes what happens in the long run, but in important examples in physics, the long run is what actually is observed.
It was Bergson, on the other hand, who most emphatically
rejected the oversimplification of experienced time by science
in its scientistic forms. It is the mixture of these two themes
that Ekeland presents to us, together with a vivid and non-
technical account of some of the major phenomena of the mathematical modeling of time in the decades since the Second
World War.

This book is not a history of this development, and I should
not advise the reader to take it as such. Too much is left out,
especially in the period between the two world wars. And how
could such a history leave out such figures as Kolmogorov and
Pontrjagin in Russia, and Marston Morse and Hassler Whitney
in the United States. Three names appear as the modern heroes
of this book, Arnold, Smale, and Thom. These are indeed
worthy heroes, but the reader should not be deceived by Ekeland's enthusiasm into believing that they form any kind of
unified school. Indeed, on one particularly controversial theme
of the book, they are hardly even on speaking terms.

This is the theme of so-called catastrophe theory. Invented
by René Thom and popularized by his disciple, Christopher
Zeeman of Warwick University, catastrophe theory has received a large amount of public attention and has generated an
enormous backlash in the mathematical world and in the sci-

entific world. It has an odd place in the mathematics of time, since catastrophe theory, almost like Plato, claims to deal with Timeless Forms, but it did indeed emerge from a close co-existence with the theory of dynamical systems. Ekeland has been criticized for being too sympathetic with catastrophe theory, and also for not being sympathetic enough. The origins of this criticism in either direction are the obvious ones. My own estimate of his assessment is that he gets it about right. However, if the reader would like a sample of the rhetoric of Professor V. I. Arnold of the University of Moscow in a vitriolic assessment of catastrophe theory in his short book *Catastrophe Theory* (published in English by Springer-Verlag in Germany [second revised edition in 1986]), let me cite two passages as edifying examples: "The beautiful results of singularity theory happily do not depend on the dark mysticism of catastrophe theory" (p. 90); "The deficiencies of this model and many similar speculations in catastrophe theory are too obvious to discuss in detail. A remark only that articles on catastrophe theory are distinguished by a sharp and catastrophic lowering of the level of demands of rigor and also of novelty of published results" (p. 9).

Similar comments (though more polite in tone) have been published by Professor Stephen Smale of the University of California at Berkeley. Indeed, the trio of Arnold, Smale, and Thom is not a happy harmony.

Ekeland has achieved the important result in this book of putting forward some of the most interesting and provocative scientific ideas of the present period with balance and precision in eloquent and vigorous prose.

FELIX E. BROWDER

INTRODUCTION

Quite some time ago, I was asked to write a book on René Thom's catastrophe theory. At the time, it was already clear to me that catastrophe theory was just one chapter among many that would describe the spectacular progress that mathematics had made in understanding time and change. The new ideas brought into mathematics had already found their way into physics. The time had come for a still wider audience to hear about them, and to be initiated in strange attractors and Feigenbaum cascades. This was the book I wanted to write. I would proclaim the revolution that these new ideas would bring to the practice of science and our conception of knowledge.

But then I found out that these new ideas were almost one hundred years old, and that this book of mine had been written several times already. Poincaré, at the turn of the century, had discovered many crucial facts, and introduced the right ideas for dealing with them. He had even gone to the trouble of explaining his mathematics to laymen in a series of popular books which remain unequaled. Bergson, too, had thrown much light on the subject, and had analyzed in depth the way exact sciences account for time. Moreover, both of them had enjoyed quite a reputation in their time, and they had been printed in respectably many thousands of copies, so one might well surmise that they had been read.

Why try again? Even my disappointment was as old as the world: "Is there any thing whereof it may be said, See, this is new? it hath been already of old time, which was before us. There is no remembrance of former things; neither shall there be any remembrance of things that are to come with those that shall come after" (Eccles. 1:10–11).

And yet I believe that there is still something to be said, and that the same old story can be told another way. New facts have been uncovered in support of Poincaré's early intuitions, new ideas have blossomed in the huge field he opened to investigation, new masters have arisen—Thom, Smale, Arnold. There is now a host of surprising facts and telling paradoxes, which can serve to illustrate this field of knowledge for the layman as a skillful photographer can convey the spirit of a strange land.

So this is the task I set myself to accomplish: to sum up, in a few pictures, the mathematics of time, which is the common background of much of contemporary science.

Successful pictures have a way of imprinting themselves on our minds and influencing all subsequent perceptions. This is as true in science as in the arts. Kepler's three laws, for instance, have been much more than an astronomical curiosity. The picture of the planets revolving around the sun in elliptical orbits has settled in the minds of generations of scientists and laymen. It is one of the constant but unspoken references of modern thought, and Newton's discoveries have set the pattern which all scientific theories strive to imitate.

Nowadays, a few striking pictures—Arnold's cat, Smale's horseshoe, Thom's cusp—are contenders for a similar influence. They have awakened an echo in all domains of science, and they will soon belong to our cultural background, like family portraits hanging on a wall, pictures so well known they cease to catch the eye, so that one has to remove them and feel the resulting vacuum to experience how important they really are.

I will try to present some of those pictures. It is not an easy task, because the pace nowadays is so quick, with new problems to be solved tumbling upon partial solutions to old ones, that anyone involved in scientific research has to run very fast just to stay in the same place. It is difficult to abstract oneself from absorption in immediate problems and try to take a broader view. But it is certainly well worth the effort. I have derived great enjoyment from writing, and then translating, this book, and I hope some of this enjoyment will be shared by the readers. May they derive a true picture of what this part of mathematics is about, and feel the thrill of understanding.

ACKNOWLEDGMENTS

I would like to thank my friends Jean-Pierre Aubin, Felix E. Browder, Jean-Marc Lévy-Leblond, and René Thom, who, each in his own way, stimulated the intellectual curiosity which lies behind this book. My thanks also go to Renée Weber for reading the English manuscript.

1

THE MUSIC OF
THE SPHERES

THE MARVEL OF KEPLER'S LAWS

The picture we see in fig. 1.1 is a familiar one. It shows a planet revolving around the sun in an elliptical orbit. As usual, the orbit is depicted as much more elliptical than it actually is, so that it may be clearly distinguished from a circle. We find this picture in elementary textbooks, along with the statement that the earth revolves around the sun. It all seems so simple and natural that it is hard to remember that it was a major scientific discovery, coming several thousand years after humanity began to study the stars.

For more advanced students, we have Kepler's three laws:

I *Planetary orbits are ellipses one focus of which is occupied by the sun.*

II *The (imaginary) line segment SP connecting the sun and the planet sweeps out equal areas in equal times.*

III *Let planet P have period T (this is the planetary year) and semimajor axis a (this is the farthest it gets from the orbit's center O). Similarly, let planet P' have period T' and semimajor axis a'. Then T^2/a^3 and T'^2/a'^3 are equal numbers.*

Kepler's first law gives the shape of the orbits. The second law determines the speed along the orbit. The farther the planet is from the sun, the longer segment SP becomes, and the slower its extremity P must move if the area it covers in one second is to remain constant. So the planet slows down when it gets farther from the sun, accelerates when it gets closer. Kepler's

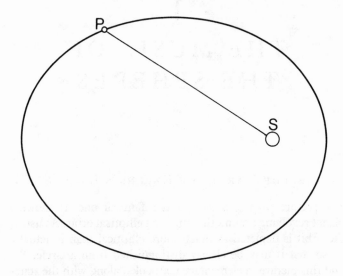

FIG. 1.1. An elliptical orbit. The sun (*S*) occupies one of the foci.

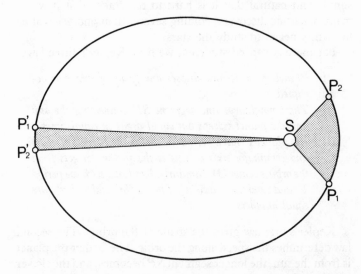

FIG. 1.2. The law of equal areas. The planet takes the same time to travel along the arcs P_1P_2 and $P_1'P_2'$ (shorter, but farther away from the sun).

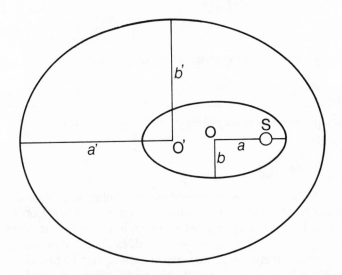

FIG. 1.3. Kepler's third law, relating the period to the dimensions of the orbit. Here $a'/a = 2$, so $T'/T = 2^{3/2} = 2.82842. \ldots$ The Keplerian ellipses must have the same focus S (at which the sun lies) but not the same center. Their shape, that is, the particular value of b/a, does not matter.

third law relates the speed of motion to the dimension of the orbits. The larger the orbit, the longer the planetary year. Pluto, for instance, is 100 times farther away from the sun than Mercury, so, by Kepler's third law, the Plutonian year is 1000 times the Mercurian year.

Together with the fact that planetary orbits all lie in practically the same plane, Kepler's three laws give a complete description of planetary movements: nine nested ellipses, from Mercury to Pluto, nine planets revolving in the same direction, like a gigantic and lopsided merry-go-round.

The year in which Kepler discovered that the orbit of Mars is an ellipse was 1605. His two first laws were published in his book *Astronomia Nova* (1609), the third one in *Harmonices Mundi* (1618). It can be stated without fear of exaggeration that they constitute the greatest scientific discovery of all time. They provide a complete and simple answer to questions which

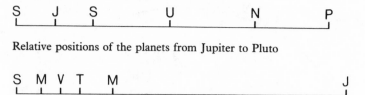

Relative positions of the planets from Jupiter to Pluto

Relative positions of the planets from Mercury to Jupiter

Fɪɢ. 1.4.

had tantalized the most distinguished scholars for centuries, Eudoxus of Cnidus, Aristarchus of Samos, Ptolemy, Copernicus. Listen to Kepler's victory hymn in the preface to *Harmonices Mundi:* "I am now enlightened, in the midst of a most desirable contemplation, eighteen months ago by the first glimmerings of dawn, three months ago by clear daylight, and just a few days ago by the sun itself. . . . The die is cast; I will write my book, and little does it matter whether it is read now or has to await posterity. It may well wait a hundred years for a reader, since God himself has waited six thousand years for someone to behold His work."

No one who has tried to trace the planet Mars in a planetarium or on a map of the sky will find Kepler's enthusiasm unreasonable. The planet is confined within a band around the so-called ecliptic, which is just the trajectory of the sun in the sky during the year. It moves generally counterclockwise (in our hemisphere), but oscillates back and forth along its mean trajectory, sometimes even changing direction and moving backward. The general effect is like a complicated dance in the sky, one step backward for every two or three steps forward, and the resulting trajectory is as full of knots as entangled fishing tackle. The trajectories of Jupiter or Saturn pose similar problems. The inferior planets, Venus and Mercury, which are closer to the sun than the earth, challenge the observer in other ways. It is no mean feat, for instance, to find out that the "morning star" and the "evening star" are but a single planet, Venus, in different positions relative to the earth and the sun. Such knowledge presupposes some type of astronomical theory.

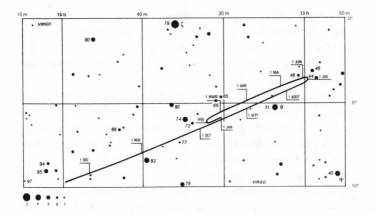

FIG. 1.5. The position of Saturn in the sky from January 1 to December 31, 1982. Notice the backward motion.

The task of unraveling the trajectories of planets in the sky has required infinitely more patience than unraveling any man-made line or rope. Kepler had at his disposal the theories—already quite accurate—of Ptolemy and Copernicus, and Tycho Brahe's very precise observations. He nevertheless had to per-form monstrous computations over a number of years. There were no electronic computers at the time, and even logarithms were a novel feature. The Pulkovo library stores thousands of manuscript pages by Kepler, covered with computations. In the *Astronomia Nova* he rounds off fifteen folio pages of com-putations by complaining to the reader of having had to do these calculations seventy times before getting the right answer.

The fisherman who tries to disentangle his lines bases his hopes of succeeding on the fact that the lines were straight and clear to begin with. There is no similar a priori knowledge underpinning Ptolemy's or Kepler's efforts—only their faith in the hidden harmony of the cosmos. Kepler's three laws vindicate generations of astronomers who, throughout history, from the Chinese to the Maya, from the Chaldeans to the Arabs, persistently tried to put order and regularity where none was apparent. To the naked eye, the movements of the planets are as regular as the flow of a river. We know that closer inspection will reveal eddies, whirlpools, a complete finer structure, within the global current. Why not be content with the overall picture, namely, a global, uniform motion of the planets along the ecliptic? Why try to explain every single deviation of a planet from its general course?

This, of course, was not "pure science," or disinterested knowledge, as we understand it today (if such a thing really exists). Since antiquity, the all-pervading influence of astrology had turned the understanding of planetary movements into a problem of the utmost practical importance. To figure out the fate of princes and potentates, it was essential to know where in the zodiac the planets (and the moon) were at the date (and hour) of their birth. To reach a long-term decision, it was crucial to know where they would be at some future date when it would take effect. Kepler himself, as court mathematician to the Holy Roman Emperor, had to cast horoscopes and draw conclusions. At the beginning of his career, he was fortunate enough to predict a cold winter, peasant uprisings, and war with the Turks, which did more for his reputation than all his later work in science. The Church, too, made great demands on the skills of astronomers. Since Easter Sunday was defined as "the first Sunday following the first full moon after the spring equinox," to find out what its date would be (or had been) in any given year was quite a complicated problem in astronomy, involving the relative motions of the earth, moon, and sun.

Apart from these practical considerations (which by now have fallen out of fashion), Kepler and his predecessors were inspired by a quest for theory, a deep conviction that there was harmony in the universe, that God had created the world with wisdom, and that the harmony of the universe, and God's wisdom, are expressed by simple, but hidden, means. We share these convictions even today, and in this we are Kepler's heirs, and followers of the long tradition which he fulfills. The belief that the secrets of nature are best expressed in mathematical language is also part of this tradition. To quote Galileo, the book of nature is written in the characters of geometry: circles, triangles, and squares.

Note that Galileo does not mention ellipses. This may seem a minor omission, but it is not. In mathematics, pictures, which support intuition, may be more important than the text, which develops and unfolds it. Ellipses are conspicuously missing, not only from Galileo's consciousness but also from that of all astronomers up to Kepler. As a mathematical tool, ellipses had been available for a very long time. Kepler refers to Apollonius (262–180 B.C.) for the basic geometric properties of ellipses, and to Archimedes (287–212 B.C.) for more elaborate ones, which he uses to show that equal areas are swept out in equal times. But until Kepler, circles remained the only geometric objects which astronomers ever considered. Motion along a circle was preferred to be uniform, that is, the speed was to be constant. All astronomical systems until the *Astronomia Nova* were to be ingenious combinations of circles and mostly uniform motions.

The simplest system goes back to Aristarchus of Samos: the sun lies at the center of the world, and the planets, including the earth, revolve around the sun with constant speed along circular orbits. This meant also that the earth had to be round and to rotate upon itself—very modern ideas indeed, considering that Aristarchus lived in the third century B.C. In addition, circles are excellent approximations to the Keplerian orbits. Of all the planets known in antiquity, Mars has the flattest orbit, and the difference between the lengths of the major and the minor axis is only 0.5%. But Aristarchus should not have placed the sun at the center of the orbits, since it really lies at a focus; in the case of Mars, it is out of position by 9%. As a consequence, by Kepler's second law, the motion is not uniform along the orbit: the planets accelerate when they approach the sun. Similar errors affect the earth's orbit. When computing the position of Mars in the sky as seen from the earth, all these errors may add up, so that at certain dates Mars will be out of position by 15 degrees. This discrepancy between the model and the observations was unacceptable, so that Aristarchus's model was rejected in favor of other constructions, less robust, but giving better results.

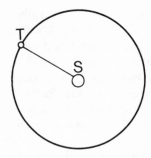

FIG. 1.6. The system of Aristarchus: the orbits are circles, with the sun at their center, and the planets move with uniform speed.

Ptolemy's system, for instance, keeps the discrepancies down to a few degrees. The most precise astronomical tables known in Kepler's time, the *Tabulae Pruthenicae* (1551) (computed using Copernicus's system), allows discrepancies of up to 4–5 degrees, as Kepler himself complains. Copernicus is reported to have set himself the goal of bringing the errors down to the same order of magnitude as the uncertainties in the observations—10 minutes of arc, one-sixth of a degree. He was not even close.

But Copernicus, and all astronomers up to the time of Kepler, did not see the problem as it was: they saw it through the eyes of their predecessors. The circle, and circular motion, were too firmly ingrained in their minds by years of training, and had shown themselves to be too successful in the past. There was no room left for any alternative. The question was no longer how to describe planetary movements, using whatever mathematical tools were available, but rather how to approximate planetary movements by sophisticated combinations of circular motions. By setting the problem in this way, the astronomers were unwittingly depriving themselves of any possibility of finding its true solution, the elliptical orbit. They had drawn a magic circle around themselves, and were searching inside the circle for something that simply wasn't there. Kepler's genius was to break the circle, reach for available tools that were lying around unused, and look in the right place.

The power of certain pictures, of certain visual representations, in the historical development of science, will be the recurrent theme of this book. It is a power, in the early stages, to initiate progress, when the ideas it conveys are still creative and successful, and it becomes, later on, a power to obstruct, when the momentum is gone and repetition of old theories prevents the emergence of new ideas.

Circular uniform motion—a point moving along a circle with constant speed—is such a picture. It lies at the heart of Ptolemy's system, which is one of the major intellectual achievements of antiquity. This basic picture comes with three variations:

1. *The epicycle.*—Imagine a small circle the center of which moves uniformly along a (fixed) larger circle. Say the center of the small circle moves uniformly along the large circle, and the small circle itself rotates with constant speed around its center. Then any point M on the small circle, when seen from the center O of the large circle, undergoes a complicated motion, which resembles very much the movement of the planets seen from the earth. It is a general drift in the direction of rotation of the large circle, with alternating phases of accel-

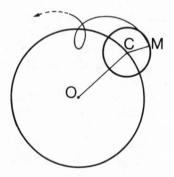

FIG. 1.7. An epicycle. The center C of the smaller circle moves uniformly along the large one. Meanwhile, the point M moves uniformly on the smaller circle. Combining these two motions results for the point M in a sequence of accelerations and decelerations.

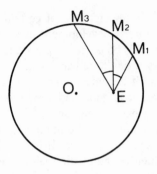

FIG. 1.8. An equant point. The point M moves along a circle with center O. Its angular speed measured from E (the equant point) must be constant, that is, equal angles (such as M_3EM_2 and M_2EM_1) are traveled in equal times (even though the arc M_2M_3 is longer than the arc M_1M_2). The farther M is from E, the faster it has to move.

eration and deceleration, sometimes even going back on its tracks.

2. *The equant.*—Take a circle, with center O, and pick a point E inside the circle, but away from the center. This will be called the *equant*. Now let a point M move on the circle in such a way that its movement, as seen from E, seems uniform. In other words, an observer located on E would see the point M move by equal angles in equal times. This would not be the case for an observer located at the center O, so the point M does not have constant speed on the circle. In fact, it will accelerate when it gets closer to E, much as in Kepler's second law.

3. *The excentric.*—For Ptolemy, the sun rotates around the earth. But the earth itself is located not at the center O of the system but at a point E', symmetric to the equant E with respect to O. This point E' will be called the *excentric*.

With the improvements made possible by the epicycle, the equant, and the excentric, the basic circular motion enabled astronomers to build models which already at the time of Ptolemy (the second century A.D.) predicted planetary positions

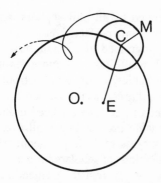

FIG. 1.9. Ptolemy's system: an epicycle combined with an equant yields a complex motion, which may be used to model the actual motion of the planets.

with discrepancies of a few degrees. No significant progress was made for over fourteen centuries, which should have shown that other methods were needed. On the contrary, the basic picture of uniform circular motion prevailed with all the more obduracy in the minds of scholars.

Good reasons were found to legitimate this persistence. Copernicus quotes Aristotle to the effect that the circular uniform motion is the most perfect and the most natural, and therefore the only one which celestial mechanics should use. Tycho Brahe arrives at the same conclusion by other means. In a letter to Kepler in 1600, he writes, "For the paths of the stars must be composed entirely with circular motions; otherwise, they would not come back perpetually and uniformly upon themselves, and would lack permanence. In addition, noncircular paths would be less simple and less regular, and would not be amenable to study or computation." As for Kepler himself, the *Astronomia Nova* bears witness to the various stages he went through. To begin with, he assigns to all planets (except the earth) circular orbits. He later adds an epicycle, and finally the ellipse comes to his mind. In his own words: "My first mistake was to believe that the planetary orbits were perfect circles. This mistake cost me a great amount of time, the more so as it had been defended by all philosophers, and was, from the metaphysical viewpoint, perfectly reasonable."[1]

Such a deeply ingrained idea would also take a lot of time to be uprooted. It would take one more century, and the help of Newtonian mechanics, to convert all astronomers to Kepler's ideas. However, in due course the astronomical revolution took place, and Kepler's three laws replaced uniform circular motions. Then the same process began all over again, with the emergence and success of a simple geometric model, which would fashion the minds of scholars for generations. The Keplerian picture too would register tremendous successes, and lead scientists to believe that they held the key of the universe. Now is the time of its decline.

CELESTIAL MECHANICS

Ptolemy's circles and Kepler's ellipses are two different ways of expressing in mathematical language a basic stipulation: we want natural phenomena to be permanent and regular, that is, predictable.

During the eighteenth century, the universe used to be compared to a clock. What model people had in mind, whether it was the latest version of Ptolemy's system or the first drafts of Newton's theory, is irrelevant. The essential point is that the comparison be made possible, so that man might feel more at home in the world. One did not hesitate to carry the comparison even further: for Voltaire, and for his contemporaries, "l'horloge implique l'horloger," "no clock without a clockmaker." Not before the end of the eighteenth century would a scientist commit himself to a deliberately atheistic position. The first one to do so may have been Laplace, when asked by Napoleon what place was left to God in the system he had built. He gave the famous answer: "Sire, I found no need for that particular assumption."

1. *Astronomia Nova*, chap. 40 (author's translations throughout, unless indicated otherwise).

But other divinities have arisen. The nineteenth century is the time when science begins to be written with a capital *S*, and acquires ascendancy over people's minds. It is a new religion, the adepts of which display the enthusiasm of new converts, and often also their intolerance and narrowness of mind. Let us hear how arrogant Laplace can be: "In these times of ignorance, mankind was a far cry from understanding that nature unflinchingly obeys permanent and unchanging laws. If the phenomena happened in regular succession, they were thought to depend on finality. If no kind of order was apparent, they were thought to depend on chance. And if they displayed some extraordinary feature, and seemed to contradict the natural order, they were considered to be tokens of divine wrath."[2]

The message is clear: there is a permanent order in nature. Phenomena must be regular, and any appearance of disorder is an illusion which will yield to deeper analysis. To be sure, at the time of Laplace, the articles of faith had been written down, the mystery of gravitation had been unveiled, and science had accomplished some of its major miracles. The belief, however, can be traced to a much earlier time. In fact, it has regulated all the development of astronomy since antiquity. If scholars kept refining the models and improving the observations, it is because they believed they could be fitted together, and hoped for a perfect match some day.

Kepler, for instance, after laboring for years on a certain model, finally rejected it because, under certain circumstances, the discrepancies between theory and observation were up to 8 minutes of arc. That is no larger than an ordinary dinner plate as seen from hundred yards' distance. It would seem to be quite a negligible angle, and the observations antiquity has bequeathed us did not even reach that level of precision. Unfortunately, in Kepler's case, the observations he had to contend with were due to Tycho Brahe. As he himself puts it, "As for us, to whom divine providence has given Tycho Brahe, an observer so outstanding that he can show the 8 minute mistake

2. *Exposition du système du monde*, book 4, chap. 4.

committed by Ptolemy, we have to accept God's gift gratefully, and benefit from it. This means that we cannot forgo the effort of discovering at last the true structure of planetary motions."[3]

The search for higher precision thus follows from the quest for a hypothetical "true structure" which would reveal, once and for all, the hidden laws of nature. Miraculously, this quest was successful, and Newton brought back the Holy Grail. The enthusiasm he aroused was proportionate to his achievements. Alexander Pope writes:

> Nature and nature's laws lay hid in night.
> God said: Let Newton be. And all was light

and Laplace ascribes to the *Principia Mathematica*, Newton's magnum opus, which appeared in 1687, "preeminence over all other productions of the human mind."

Newton's work is all the more impressive if one remembers that he obtained his main results in 1666, at twenty-four years of age, while staying in the countryside to avoid the great plague epidemic which had broken out in London, and kept his results unpublished for many years; it is not clear why.

The title he chose, "Mathematical Principles of Natural Philosophy," is quite suggestive. Newton is not content with a description of the system's behavior, as seen from the outside. He wants to understand its inner workings, to see how it functions from the inside. Kepler's three laws describe the planetary motions, and one can use them to predict future positions up to a certain level of precision. But neither Kepler, nor anyone until Newton, had been able to answer the question, "What makes the planets move?"

In a way, Newton doesn't answer the question either. The well-known law of gravitation tells us that the planets are acted upon by the sun, and shows how this action results in the Keplerian orbits. But it does not tell us how the sun acts upon the planets. Indeed, "matter attracts matter," and the force of attraction is inversely proportional to the square of distances. But this statement raises more questions than it answers: What

3. Quoted by Koyré, *La Révolution astronomique*, p. 179.

is matter? Why this attraction? How can a force be exerted from a distance, even across vast stretches of vacuum? These questions are meaningful even today, but they were particularly pressing in Newton's time. In the physics of the day, force was exerted upon contact and was transmitted through shocks. Up to the time of Newton, there was no accepted example of force exerted from a distance.

Newton himself considered gravitational attraction more as a mathematical convenience than as a physical reality. His followers had no such misgivings, and posited "Newton's law of attraction" as the underlying basis of all our understanding of the physical world. Later in the nineteenth century, science discovered electromagnetic forces and Coulomb's law of attraction, which regulates the attraction or repulsion between electric charges in much the same way that Newton's law regulates the attraction between masses. But this time Faraday, and later Maxwell, went beyond the mere empirical fact of forces acting from a distance, and introduced the electromagnetic field to mediate this action. This field propagates at a finite (but high) velocity, so that the action is not instantaneous, in contrast to Newtonian attraction. Of course, nowadays field theory has permeated the whole of physics, including gravitation, and scientists are setting up experiments to detect gravitational waves, as in the nineteenth century they were finding out about electromagnetic waves. Einstein's theory of general relativity shows that gravitational attraction is simply the way we perceive the curvature of four-dimensional space-time. Each mass creates curvature in its immediate neighborhood, and the global geometry of space-time somehow reflects the curvature at every point. In other words, what we see as a force acting from a distance results from myriad local interactions which we do not see. Newton's qualms are vindicated, and Newtonian physics stand today as a very useful approximation, very precise in its own range of applications, but devoid of internal justification, a purely mathematical model.

But in its heyday it was believed to be much more: the key to understanding the universe. Newton himself proves Kepler's three laws from the law of gravitation, and explains the tides and the precession of equinoxes by the attraction of the sun and the moon. There are the first results in a new science, celestial mechanics, which will attract the greatest names in mathematics, Euler, Lagrange, Laplace, Poincaré, Siegel. During more then a century, its spectacular successes will set an example which all the other sciences will try to follow, and which will fill with awe the minds of many.

Strangely enough, when one actually studies celestial mechanics, the first thing one learns are that Kepler's three laws are false. To put it more exactly, they are only good approximations. The sun's attraction by itself would put each planet in its Keplerian orbit. But the sun is not the only gravitating mass: there are other planets, including a very large one, Jupiter. Their combined attraction will draw each planet away from its Keplerian orbit. These deviations, however, are amenable to computations, and astronomers very quickly developed mathematical methods to predict the position of any planet at a given date with a prescribed precision. This is known as the theory of perturbations, and its development was highlighted by two remarkable works which had a lasting influence, Laplace's *Mécanique céleste* (1798–1825) and Poincaré's *Méthodes nouvelles de la mécanique céleste* (1892–1899). To give some idea of the precision that can be achieved, let us note that the position of the planet Mercury can be pinpointed within a few miles several months in advance. And let us not forget the *Apollo* missions, or the spacecraft that were sent to the other planets. All these feats are outcomes of perturbation theory.

If the positions the planets occupy today are known, together with their velocities, it is possible by these methods to compute their positions and velocities at any future date. It is also possible to turn back the clock, and to find out the positions and velocities of the planets at any past date. Astronomers predict the eclipses of the sun and the moon, and they also figure out the exact dates of ancient eclipses recorded in historical documents. In other words, the past and the future of the solar system are entirely contained in its present. To know the state of the universe at any past or future date (the mathematics sees

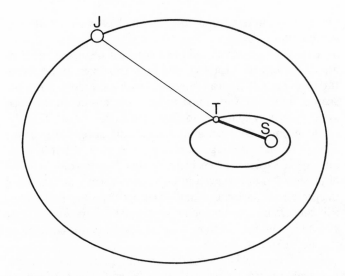

FIG. 1.10. Perturbations in celestial mechanics: planet T (the earth) is attracted not only by the sun (S), but by the large planet Jupiter (J) If J were not present, the orbit of T would be elliptical, and all of Kepler's laws would hold. In the real world, the attraction of J draws T away from the elliptical orbit. Kepler's laws are no longer exact: they are approximations, valid because the mass of J, and of the other perturbing objects within the solar system, is much smaller than the mass of the sun.

no difference), it is sufficient to know its present state accurately enough and to have adequate computing power.

In this way, the whole of time, from infinite past to infinite future, is swallowed up in the present, that ever-vanishing gap which separates what just was from what is yet to be. Past and future are seen as equivalent, since both can be read from the present. Mathematics travels back in time as easily as a wanderer walks up a frozen river.

This was the picture of time that emerged from Newtonian physics, implausible as it may seem. The scientists of the nineteenth century believed that they could encompass the beginning and the end of time, and that nothing, not even the future of humanity, was beyond their reach. It was all a matter of computation.

They even predicted the future of their own science. Here are the tasks which Laplace sets for astronomers of later centuries: to register stars and nebulae, with their movements and shapes, and their variations: to find new objects in the solar systems, mainly comets, and their orbits. There is nothing else left for them to do, since all the interesting things have already been done. As Lagrange puts it, "since there is but a single universe to explain, no one can redo what Newton did, happiest of mortals."[4] Laplace even indulges in analyzing the causes of the progress of astronomy in the future. As he sees it, progress will come from improvements in our ways of measuring time and angles and from better optical instruments, the latter being the most likely bet.

One may well be astounded by this narrow-mindedness. Laplace's program is enough to stifle with boredom several generations of astronomers. How different things turned out to be: spectroscopes, radio telescopes, lasers were the instruments of the future, and they found quasars and black holes. But Laplace is typical of his time, and for a century people lived in that kind of world, closed and smothering, where everything was known beforehand. It is in this environment that the philosophy of the Enlightenment developed, and that many political, economic, and social doctrines were born, which today are still in our way. It is also the time when explaining became separated from understanding. The law of gravitation provided a mathematical model which enabled a few experts, using long and impenetrable calculations, to predict any astronomical situation with great accuracy. But no one knew what the gravitational force was, or how it could propagate through a vacuum, crossing enormous distances instantaneously. This was the time when scientific thought separated from natural intuition, and quantitative appraisal from qualitative understanding.

The new doctrine was not to be resisted. It made up for its philosophical weaknesses by its tremendous efficiency. New

4. Quoted by Koyré, *La Révolution astronomique*, p. 163.

successes kept stimulating the zeal of its practitioners. Let us recall, for instance, the discovery of the planet Neptune. Irregularities had been found in the motion of Uranus and were attributed to the attraction of an unknown planet. Le Verrier in Paris and Adams in Cambridge separately undertook the necessary calculations, and after a few year's work, in September 1846, Le Verrier was able to write to a colleague in Berlin, asking him to look at some particular spot in the sky. Sure enough, Neptune was there. An astronomer had discovered a new planet without even raising his head.

It was a tremendous success. Failure followed some years later, when Le Verrier applied the same methods to the irregularities in the motions of Mercury and announced the discovery of a new planet, Vulcan, which never showed up. Nowadays, we know that these irregularities come from general relativity and cannot be explained in the Newtonian model. But another success was scored in January 1930, when Pluto was discovered in circumstances somewhat similar to those of the discovery of Neptune. And just recently I learned that the mass of Pluto is insufficient to account for the observed perturbations in the orbit of Neptune, so that there may be still another massive object beyond Pluto, a planet or a degenerate star.

So the war cry of astronomers in the nineteenth century is still heard today. "Give me but pencil and paper, and I shall reconstruct the world!"

CLASSICAL DETERMINISM

These great ambitions were codified very early. In the first edition of the *Principia*—a collector's item—Newton lays down two rules, which are still in effect today:

1. No more causes should be called upon to explain natural phenomena than those that are both true and sufficient to that effect, for nature is simple, nor does it provide causes in excess.

2. This is why the causes of natural effects of the same kind are the same.

These rules will be greatly refined later on, to yield classical determinism, but one already recognizes the linear filiation from cause to effect, so well adapted to the physical sciences and so inadequate for biology and the social sciences. For everything that will happen tomorrow a cause can be found today, and knowing the cause sufficiently well will enable us to predict the effect. Chance itself can be accounted for: it is seen as an imperfect kind of knowledge, which we have to resort to when a phenomenon has too many causes for us to keep track of, or when they are too small to be adequately observed. As Einstein put it, "Gott würfelt nicht." God does not play dice, and there is always the hope that in the future, deeper theoretical understanding and better computing methods will reveal the hidden determinism of apparently random phenomena. Meanwhile, probability theory and statistical methods provide more than adequate information.

The most perfect mathematical expression of determinism is the differential equation. It is a mathematical tool which was created by Newton to derive Kepler's laws from the gravitational pull. Even today, a deterministic system is a system which can be modeled by a differential equation. It has the

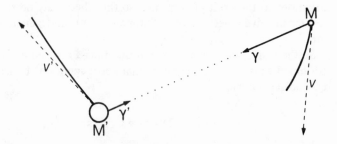

FIG. 1.11. Newton's equations. Two objects M and M', with masses m and m', travel with velocities v and v'. If they did not attract each other, they would travel in a straight line with constant velocity. According to Newton's law, they do attract each other, and the force F exerted on each of them is the same. But this force gives rise to different accelerations, since the masses are different: $\gamma = F/m$ and $\gamma' = F/m'$. Since $m' > m$, we will have $\gamma' < \gamma$, and the trajectory of the more massive object M' will not be as curved as that of M.

fundamental property that its state at any time completely determines all subsequent (and preceding) states, and there are numerical procedures to compute the state at any time t from the initial state and the equation.

A differential equation is an instantaneous relation between position, velocity, and acceleration, which is supposed to hold throughout the motion. Integrating, or solving, the equation, means to deduce from this relation the actual trajectory which will be followed and the motion thereon.

Let me first try to put this idea across for literary-minded people. As any reader of Alexandre Dumas's *The Three Musketeers* will remember, d'Artagnan first met his servant Planchet on the Tournelle bridge, while the latter was spitting down into the Seine and observing the rings on the water. D'Artagnan's colleagues claimed that this was definite evidence of a thoughtful and contemplative mind, and on their advice d'Artagnan hired Planchet.

It is indeed fascinating to watch the river from a bridge. One can see flow lines, which are mostly permanent, and eddies, which usually move around. The water even seems stationary at times, with a smooth surface on which the flow lines hardly appear. How tempting it is then to reveal the hidden movement by dropping some light object on the water, and watching it float swiftly out of sight along a flow line! One may even experiment, by dropping two objects in succession at the same place, and checking that they follow exactly the same trajectory.

One then starts looking at the surface of the water, and imagining at every point a little arrow giving the direction and the velocity of the current at the point. From such a picture, it would be easy to find the trajectory of an object dropped at any specified spot. This is precisely what differential equations are about. The equation itself is that specification of the velocity at every point. Integrating the equation means finding these trajectories.

Another, less poetic way to approach differential equations is through computations. Let us imagine a point moving on a straight line, away from a prescribed origin, in such a way that

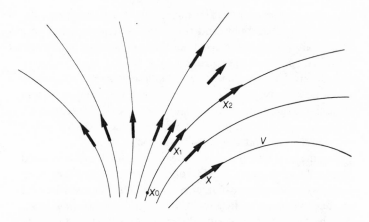

FIG. 1.12. A first-order differential equation is an instantaneous
relation between the position x of an object and its velocity v. If the
initial position x_0 is known, the subsequent positions x_1 (at time 1),
x_2 (at time 2), x_3 (at time 3), and so forth, are completely
determined. By plotting them, one obtains a solution of the
equation. Imagine little arrows indicating the direction of the
current in a stream: they give the velocity of the water at every
point, that is, they define a differential equation. If a small object
falls into the water, it will be carried away by the current in the
direction indicated by the arrows. The point of impact would be x_0,
the initial position, while the actual trajectory followed by the
object would be the solution starting from x_0.

its speed v at every instant is inversely proportional to its distance x to the origin, $v = 1/x$, say. This is a very nice one-
dimensional differential equation (the preceding one was two-
dimensional).

To integrate this equation, that is, to find the corresponding
motion, a computer will proceed as follows. It first asks for
the initial position, that is, the position at time $t = 0$ when
the clock starts running. It is provided with the answer, $x_0 = 2$,
say. It then computes the speed from the equation: it is $v = 0.5$.
Keeping this speed constant during the next time interval, from
$t = 0$ to $t = 1$, so that the length traveled is 0.5, it will then
compute the new position, $x_1 = 2 + 0.5 = 2.5$. It will then
compute the new speed at time $t = 1$, namely, $1/2.5 = 0.4$,

keep it constant up to time $t = 2$, and thereby find the new position $x_2 = 2.5 + 0.4 = 2.9$. In this way, the successive positions at times $t = 0, 1, 2, 3, \ldots$ will be computed.

Clearly, this is just an approximation to the true motion, because the equation $v = 1/x$ tells us that the speed is not really constant on time intervals but decreases slightly. Better approximations can be computed by using smaller time intervals, of length 0.1 instead of 1, or better still by using intervals of length 0.01. The true solution is actually known: the position x at time t is given by $x = \sqrt{2t+4}$. For time $t = 1$, for instance, we get 2.45 instead of 2.5, which shows that even our first approximation was serviceable. The important thing to remember is that an instantaneous relation between position and speed will enable us to compute both for all times, provided that the initial position, at time $t = 0$, is known.

Newton's work in the *Principia* provides the first and most perfect example. The law of attraction and the fundamental relation of dynamics, acceleration = force/mass, give a differential equation for planetary motion. Newton formulated this equation and integrated it, thereby finding Kepler's three laws.

His work cannot be valued too highly. To derive Kepler's law from gravitational pull, he had to create a new mathematical technique, calculus, capable of formulating and handling differential equations. The technical and conceptual difficulties to be overcome were tremendous. How does one define instantaneous velocity? What is meant by "velocity at time t," since an instant of time has no duration, and there will be no actual motion during time t? And how can an instantaneous relation between position and velocity be integrated into a global solution? All this is calculus, taught today in college freshman courses and even in high schools. Together with this technique, freshmen will have been taught that the solution of a differential equation is entirely determined by the initial state. Thus, the idea that the state of the world today contains its past and its future will have been passed on in the guise of a theorem in mathematics.

From then on, any user of differential equations, and there are many of them, since mathematics has no other way to model time, will encompass eternity in one instant of time. The sci-

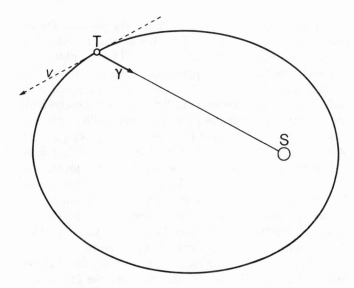

FIG. 1.13. Integrating Newton's equations: the sun (*S*) is so massive that its own gravitational motion, under the attraction of the planets, can be neglected. It is therefore considered to be immobile, while the planets move in its gravitational field. If the sun were not present, planet *T* would escape in a straight line with constant velocity. The gravitational field of the sun bends this straight line and accelerates the motion. The precise calculations, made possible by Newton's laws, show that the resulting orbit must be an ellipse, and vindicates all three Kepler's laws.

entist who writes down a differential equation to represent a physical system has on hand all its future evolution, provided only that the initial state can be observed with enough accuracy.

Better still, Newton's solution of Kepler's problem, the first example and greatest success of this technique, will lead scientists to believe that differential equations lead to stable and regular motions. This belief comes easily when Kepler's laws are taught uncritically as basic facts, and is strengthened by teaching and experience. This is because education depends very much on tradition, even in mathematics: one teaches only what is well understood and has proved its usefulness, and skips unpleasant facts and unknown areas.

The fact that the solution of a differential equation is completely determined by the initial data, that is, that position and velocity at every future or past time depend only on position and velocity at time $t = 0$, is mathematically well established. But what are its practical implications? Does it mean that the motion has to be regular, or is it compatible with some kind of randomness? This is almost never discussed. On the other hand, one discusses at length numerous examples of carrying the computations down to the end, where the result is indeed some kind of regular motion. A young scientist who has been through the conventional course of education will try to come up with models showing the same pattern of regularity, thereby increasing the stock of examples and strengthening the consensus within the community that this is the right outlook.

In science, as everywhere else, there are few true creators, people able to leave the beaten track and to come up with new ideas. It is very tempting to deem a problem interesting because half the people you know are working on it. But truly deep and difficult problems promise no easy returns, and do not attract people eager to publish. Poincaré makes a distinction between problems that nature sets up and problems that one sets up. And it is Poincaré who will initiate the critical analysis of classical determinism, thereby opening the modern era. His tremendous powers of analysis were directed toward the most formidable stronghold of Newtonian physics, celestial mechanics, with devastating effect.

2

THE SHATTERED CRYSTAL

THE IMPOSSIBLE CALCULATIONS

As celestial mechanics went from one portentous success to another, a particular view of the world, known today as scientism, evolved and became consolidated. By the end of the nineteenth century, it was well entrenched, not only within the scientific community but also among philosophers and the public at large. It held that the laws of nature were known, or would be known, and that the world was deterministic, so that predicting the future from the present was just a matter of computation. Whoever could perform these calculations in some particular case would be assured of immediate glory: Lalande, who predicted the date when Halley's comet would return; Le Verrier, who discovered Neptune. People were looking forward to the not so distant day when such achievements would become commonplace, and when the world would be an open book before our eyes. So important had science become that priority quarrels became matters of national pride. The question of who discovered Neptune, Le Verrier in Paris or Adams in Cambridge, remains a touchy subject between the French and the English.

There is, indeed, some reason for pride. Scientists could discover new worlds through computations, without even lifting their eyes from the paper they were scribbling on. Before the Newtonian era, such a feat would have been inconceivable. To take the nearest example, when Columbus discovered the New World, he left Europe with three ships on a dangerous voyage, and when he set foot in America he believed he was in the Indies. His was a chance discovery, a stroke of luck, whereas Le Verrier and Adams found exactly what they were looking for.

Science carries an aura of prestige, and the absentminded scientist has found himself a place in the popular imagination: a funny little man, always lost in some complex calculation and oblivious of his surroundings, almost helpless in daily life, and slightly ridiculous. For all his failings, he is surrounded with respect and is in fact a disquieting figure, for one never knows what deep insight or deadly invention will spring up from his unuttered thoughts. He wields the power of science, this almost magical power which has built the society we live in. We too, in our school days, have held a small fraction of this power in our hands, for we also have learned to calculate. Of course, we are neither Newton nor Le Verrier, but we have held the tool they were so proficient with, and perhaps, with a little luck and a lot more work, we could have done very well too. Even if scientific successes evade us, school has shown us the basic tools of the trade, and we may sense a continuity between what we learned and the ever-expanding frontiers of knowledge. So let us join in the famous battle cry, which contains the gist of former achievements and expresses the hope for more advances: "I believe that two and two make four, and that four and four make eight."

This is the situation toward the end of the nineteenth century. Unfortunately, the apple is rotten at the core. Science has striven for over a century to build an imperishable temple on the foundations Newton laid, and the construction is dazzling indeed. But the columns are already cracked; as a matter of fact, they were cracked from the very beginning, and they will soon fall and bring down the whole building. These cracks are there to be seen: one needs a really close look, but they certainly can be detected even if there is no saving the construction. But the wardens of the temple keep the bad news to themselves and try to paint everything over. The tourists keep visiting. They will not be told until the construction finally collapses.

They should have guessed long before! There was clearly something amiss. To begin with, all these computations were long, so long that very few people were willing to undertake them, and there was no question of checking them, let alone double-checking. Lalande, in his search for Halley's comet, started computing in June 1757, and announced his results in November 1758. It took Adams two years, Le Verrier one year, to discover Neptune. No one was going to spend one or two years of his life just to check their calculations.

Worse still, these computations were often wrong—or shall we rather say, off course by a sizable amount? Halley's comet reached perihelion one month before the date predicted by Clairaut and Lalande. Compared with 75 years (Keplerian period of the comet), or with 618 days (the actual value of the perturbation they set out to compute), one month off does not seem too much. Still, it shows that there are limits to computational power. Adams and Le Verrier announce two different orbits for Neptune: Le Verrier's is the better one, happily for the French, but it is not really good. He predicts a mean distance from the sun of 35–38 times the radius of the earth's orbit, and a period (Neptunian year) of 207–233 years, whereas the true values are 30 and 164. Ill-wishers even point out that, had Le Verrier performed the same computations forty years earlier or later (one-fourth of a Neptunian year), the discrepancy between the predicted position and the actual one would have been so great that it would have been practically impossible to find the planet in the sky. Flammarion's *Astronomie populaire*, a perennial of French literature, and one of my favorite books when I was a child, says it with Victorian undertones: "These large discrepancies sowed doubts in peoples' minds."

These minds were quickly put back on the right track. The most damning facts were concealed, to avoid creating further embarrassment. How, for instance, did Adams and Le Verrier set up their computations? They needed to know the mass of the unknown planet which was perturbing Uranus away from its Keplerian orbit. They did not compute it; they just guessed. That is, they put in what they thought would be a likely mass for a planet they knew nothing about. This was also the reason why their computations were so far off target: they could not be any better than the initial guess. Adams had chosen the mass of Neptune to be 45 times the earth's, and Le Verrier chose 32, whereas the true value is only 17. All these impressive computations are just Noah's cloak cast on the initial guess.

One cannot but wonder that so many distinguished scientists, mathematicians or astronomers, never seemed to wonder why the actual computations of perturbation theory were so difficult and unreliable. Instead, they unceasingly preached a gospel announcing that the law of gravitation, and a few others like it, would enable science to explain and predict everything. As far as one can see, this claim may be well founded in theory, but it has little practical relevance. To put it into practice, one must be able to perform the necessary calculations, which are usually out of our reach, even in this era of computers.

Perturbation theory itself is not very difficult. If, for instance, it is the actual orbit of the earth one is after, one would use the Keplerian orbit as a starting point—a first approximation, in scientific jargon. If the Keplerian orbit is not good enough for whatever purpose the scientist has in mind, he will look for a better one, closer to reality—a second approximation. To get it, he will take into account the attraction of the most massive planet, Jupiter. Its effect on the motion of the earth will be computed by using two simplifications:

1. The orbit of Jupiter will be considered Keplerian, which it is not, because of various perturbations, including those caused by the attraction of the earth and other planets. All these perturbations will be neglected.

2. In computing the effect of Jupiter's attraction on the earth's motion, one will not use the actual (Newtonian) equations of motion, but simpler ones, called the linearized equations. The procedure by which one obtains the linearized equations from the true equations is very similar to the procedure by which one replaces a curve in the plane by its tangent at some point. This approximation is valid if one stays close enough to the chosen point. In a similar way, the linearized equations are valid as long as the earth does not stray too far away from its Keplerian orbit.

If the orbit obtained in this way is still not good enough, one will then look for a better one—a third approximation, taking more perturbations into account, those caused by the attraction of other planets and of the moon. If still better approximations are needed, one might call upon other perturbations—those due to the imperfect sphericity of the earth, for instance. There are a great many possible perturbations, and one includes only those that are necessary to achieve the prescribed accuracy.

Perturbation theory is simple, but the actual computations are tremendous. As a matter of fact, they very quickly become intractable. To be sure, the nineteenth century has scored some brilliant successes, and we are much better off nowadays, thanks to computers. A very long computation, much longer than Lalande or Le Verrier would ever have considered, now takes at most a few hours. But there still is a limit to what one can do, and it remains surprisingly close.

Suppose, for instance, that one is dealing with a system of heavenly bodies, moving together under the influence of gravitational forces, and one is interested in solving the equations of motion. Say that there are N bodies in the system, and they have similar masses, so that one cannot use perturbation theory. Then there will be exactly $N(N-1)/2$ interaction terms in the equations, one for each pair: 3 if there are 3 bodies, 45 if there are 10, 4950 if there are 100, and 499,500 if there are 1000. Since none of these interactions can be neglected, all must be kept track of, and their number provides a good measure of the complexity of the problem. In other words, if you increase the number of bodies by a factor of 10, you increase the complexity of the problem by a factor of about 100. So, if one is ever able to build computers which will be one million times as powerful as present-day ones, then one will be able to solve problems with one thousand more bodies, not one million. This has to be seen in the light of two background facts. The first is that the complexities of the N-body problem are such that one is content nowadays with getting numerical solutions for $N = 3$. The second is that our Galaxy contains about one hundred billion stars, all interacting under gravitational forces.

The situation becomes even worse if, instead of increasing the number of bodies in the system, one merely tries to improve the accuracy of the predictions. No computation can pinpoint the position of a planet for all future (or past) times. What can be done, given some margin of error E and some finite horizon T, is to compute the position for all times less than T, with an error less than E. A historian, for instance, tracing eclipses in the past, would typically be interested in a period ranging from now to classical antiquity, about 3000 years, and would accept errors of several hours (as long as they don't change the date). On a lunar mission, on the other hand, the time span T is much smaller, two or three days, but one would want to locate the spaceship with great accuracy, within 100 meters or so. These numbers T and E, time span and margin for error, must be agreed upon beforehand, and all the computations rely on them. The greater T is, and the smaller E, the more difficult the computations.

The program basically divides three-dimensional space into cubes with side E (and volume E^3), and the computations will locate the spaceship, or whatever object is being traced, within one of these cubes. If greater accuracy is desired, a finer mesh must be used, with smaller cubes. If, for instance, one wants to improve accuracy tenfold, one must use cubes of size $E/10$ instead of E. There are $10^3 = 1000$ such small cubes in a single large one, all of which must be allocated some place in the memory of the computer. This means that the computations will become 1000 times more complicated! In other words, a program which can trace a lunar mission within 100 meters in 1 second will run for almost 20 minutes to pinpoint the same trajectory within 10 meters. By this time, of course, the result will be meaningless, the vessel having long since left the computed position. Alternatively, if one still wants the answer within 1 second, then some way must be found to perform the same computations 1000 times faster, which clearly cannot be achieved by technological means only. From one generation of computers to the next, speed increases tenfold, not a thousand fold.

In this century, the development of digital computers has vastly expanded the range and accuracy of predictions in celestial mechanics. But the frontier of feasibility is still there, even though it has been pushed farther away. There are a great many computations that cannot be performed now or in any foreseeable future.

Here, for instance, is a question which has been around for three hundred years: how did Saturn get its rings, and how is their structure to be explained? From the earth, there appear to be several of them, with varying degrees of brightness, all lying in the same plane, and separated by darker bands, the most important of which is the so-called Cassini divide. It has been known for a long time that the rings are not solid, but are made up of a multitude of independent particles, orbiting together around Saturn. It is also known that gravitation is not the only force acting on them. Shocks between particles play an important role in explaining, for instance, why the rings are flat. But why several rings? And why these gaps between them?

The question has been very much in dispute. Most experts attribute the gaps to gravitational perturbations exerted by the heavier satellites of Saturn, but there is, as of today, no satisfactory theory to account for the observations. The *Voyager* missions have shown the situation to be even worse than expected. On closeup, the three traditional wide rings dissolve into hundreds of narrow rings in all possible shapes, flat or braided, and the so-called gaps are jammed with particles of all sizes, rocks a few inches across and asteroids a few miles wide—everything but the kitchen sink. No one has a theory for that.

But even if we have no theory, we can still try some computations. The computers will beat the astronomers at their own game. After all, Saturn's rings should be an outcome of Newton's law of gravitation. If we start from a flat cloud of particles, evenly distributed around Saturn, and let it evolve under the influence of Saturn and its moons, we should see its shape change slowly and the ring structure develop. It is probably impossible to conduct this kind of experiment—unless some appropriate small-scale physical model is found—but we can at least find a numerical model and run it on a computer. With present-day graphic capabilities, one might even make a movie out of it: *How Saturn Got Its Rings*—a popular sequel to Kipling's *How the Leopard Got His Spots*.

There is a long way to go before cashing in on this box office success. Several numerical models have been tried on larger and larger computers, and none has ever run long enough to

FIG. 2.1. The rings of Saturn, photographed by *Voyager 2* from a distance of 9 million kilometers (August 17, 1981). Photo, Public Information Office, Jet Propulsion Laboratory, NASA.

show any hint of ring formation. Of course, programs do not run forever; computer time is very expensive, especially on big machines, and the allotted budget can be spent very quickly. But the fact remains that the genesis of Saturn's ring system is out of reach of present-day computing facilities, even though it should be a simple consequence of the laws of celestial mechanics.

One can of course argue that Saturn's rings, as we see them today, are the outcome of a process extending over billions of years, and that this span of time may just be too long for present-day computers, but eventually computers will become powerful enough to solve this problem and many others. Unfortunately, as we pointed out before, there is no guarantee that it will ever be so. It may be that, to perform the computations efficiently, we need computers which would be one million times as fast as the ones we use today, and because of physical limitations, it is not clear that such computers will ever be built. In other words, to understand Saturn's rings, we should rely on theoretical advances at least as much as technological progress.

It is always a matter of surprise to me how little is known in celestial mechanics. The most basic questions have gone unanswered since Newton's time. What is the earth's true trajectory? Is it gradually nearing the sun, and will it finally be swallowed up? Or is it slowly drifting away, finally to escape into the cosmic vacuum? No one knows. The Keplerian orbit is but an approximation, good enough to describe the trajectories over a few years. Perturbation theory, with the help of modern computing facilities, extends this span to a few thousands of years. This is a lot by human standards, since it covers all recorded history, but very little by astronomical standards. The past and the future of the solar system are hidden from us.

POINCARÉ'S CONTRIBUTION

Henri Poincaré (1854–1912) was a mathematical genius. His contributions range over all fields of mathematics, but celestial mechanics held a special fascination for him, and he returned to it over and over again, leaving a lasting imprint on the field and founding the modern theory of dynamical systems. Even in his lifetime, his work in celestial mechanics won him the highest honor, since his 1889 memoir *Sur le problème des trois corps et les équations de la dynamique* was awarded a special prize

created by King Oscar II of Sweden. There were no Nobel
Prizes at that time, and this award attracted much attention.
His later work, three volumes entitled *Les Méthodes nouvelles
de la mécanique céleste,* published between 1892 and 1899, belongs to the classics of mathematics.

Celestial mechanics was particularly suited to the tools Poincaré had developed to tackle other problems, such as the geometry of surfaces or the classification of differential equations.
As a general rule, Poincaré, who is without peer where computing is concerned, pushes his calculations as far as they will
go. When the limit is reached, he first surveys the road he has
covered thus far, and then he tries to peer ahead into the mist.
There are no mileposts or signs to be seen; only the shapes of
mountains emerge from the fog.

At this frontier of knowledge, one must change instruments.
For quantitative methods, accurate but limited in scope, we
must substitute qualitative methods, which have greater range
but less precision. Poincaré was the undisputed master of qualitative methods, which he introduced into mathematics under
the name of *analysis situs*—nowadays topology. He will remain
in history as the most penetrating critic of quantitative methods, and the great proponent of qualitative ones. Even the title
of his book carries the message: if indeed there are new methods
in celestial mechanics, who cares about the old ones?

Poincaré's criticism—even though he might not have wished
to carry it that far—is aimed at the very idea that a quantitative
model, accurate as it may be, can be used to predict the future.
This goes directly to the root of the deterministic faith, and it
is understandable that Poincaré, at the time, did not want to
draw all the conclusions of his criticism.

This is the reason why he confines his criticism to the narrow
field of celestial mechanics and hides it in a technical formulation. He states blandly, as if it were the most innocuous fact
in the world, that the equations of dynamics are not completely
integrable, and that the series which are commonly used to
provide approximate solutions are all divergent.

To understand what this means, we should first ask ourselves
what we would call a complete solution—of the three-body

problem, for instance. Given three point masses, with known positions and velocities at the initial time 0, the problem is to compute their positions and velocities at any given future (or past) time t. One hopes for a mathematical formula, depending on t and the initial conditions; substituting for t the given time, the formula would give the positions and velocities at that time. For instance, the formula $x = \sin t$ completely determines x in terms of t; if I am curious to know the sine of 10, I just take my pocket calculator, punch in 10, push the sine button, and get -0.54402111. It is this kind of relation we are looking for, slightly complicated by the fact that we need nine numbers, instead of one, to describe the positions of three points in space. A complete solution of the three-body problem would thus be a set of nine relations $x_1 = f_1(t), \ldots, x_9 = f_9(t)$, giving the positions of the three point masses by simply substituting the desired value for t.

Poincaré's statement is that there is no such thing. It is true that there is a well-defined relation between time and position, which completely determines the latter once the former is given. It is also true that, if one were able to reproduce *exactly* the same initial conditions, one would observe *exactly* the same motion, that is, identical positions at identical times. What Poincaré denies is the possibility for us to actually write down this relation in a way suitable for computations. Some steps can be taken in that direction: one can write down relations which are valid within a certain accuracy up to a certain time; but it is impossible to write down *the* relation which is valid with infinite accuracy for all times.

In other words, Poincaré has shown that it is impossible to compute explicitly the general solution of the three-body problem. This may seem strange to all those who remember t^2, $1/t$, $\sin t$, $\cos t$, e^t, and other functions from their high-school days. One knew very well what they were like for large t, and their values could be computed as far as need arose. Unfortunately, such functions are very rare; they are the ones you find on pocket calculators. One may go a bit further and learn about the so-called special functions, but here again the list is very short.

Poincaré's first result states that, in the three-body problem, the relation between time and positions cannot be expressed in terms of functions in the short list above: polynomials, rational fractions, exponentials, special functions. This first negative result is not yet decisive, for it leaves open the possibility that some other formulation might be found. The functions on the short list have nothing magical: it just happens that we have simple and efficient formulas to compute their values. For instance:

$$e^t = 1 + t + t^2/2 + t^3/6 + t^4/24 + t^5/120 + \ldots,$$
$$\cos t = 1 - t^2/2 + t^4/24 - t^6/720 + \ldots.$$

The infinite sums on the right-hand side are called series. They are referred to as convergent to express the fact that they can actually be used to compute the left-hand side. For instance, the first three terms $1 + 1 + 0.5 = 2.5$ already provide a decent approximation to $e^1 = e = 2.718\ldots$, and one obtains better and better approximations by including more and more terms in the sum.

One might therefore try a shortcut. Instead of expressing the complete solution of the three-body problem in terms of the functions on the short list, why not express it directly as a series, analogous to the preceding ones:

$$x = a_0 + a_1 t + a_2 t^2 + a_3 t^3 + \ldots,$$

where the coefficients a_0, a_1, a_2, . . . will be determined inductively in order to satisfy the equations of motion.

Poincaré's second negative result closes this loophole. He states that all series obtained in this way are divergent, which means that the sum on the right-hand side grows to infinity as more and more terms are included. A typical example of such a series is the infinite sum whose terms are all unity, i.e., $1 + 1 + 1 + \ldots$.

These series cannot be used to compute explicitly a complete solution for the three-body problem. They are, however, used to compute solutions which will be valid within some finite accuracy for some finite amount of time. They have been subject to countless studies, and Poincaré himself has made a systematic investigation of their possibilities. In his own words, the divergence of these series "is of little consequence for the time being, since we are sure that computing the first terms provides a very satisfactory approximation. One must, however, keep in mind that these series cannot provide infinite accuracy. The time will come when they too will be deemed insufficient. In addition, some theoretical conclusions one would be tempted to draw from these series are unwarranted, because they diverge. For instance, they are useless to settle the question of stability for the solar system."[1]

Poincaré has shown that, in the very midst of the most accurate and ambitious mathematical model, the Newtonian system, there is still ample room for the unpredictable. There will always be problems beyond any computation, questions that cannot be answered. The future of the solar system is probably one of these. But mathematics continues, even when computations have to stop. Using new methods of investigation, qualitative instead of quantitative, we shall try to outline the possibilities that the future holds in store, instead of seeking exact predictions in all circumstances.

Before we go into this, I would like to come back to Poincaré's work as a critic of determinism. There are two sides to it. In the first one, as we just saw, he shows that certain events in the physical world are unpredictable, even in the Newtonian model, because the necessary computations cannot be performed. In the second one, he shows something even more surprising: some events, which the mathematical model predicts, will not happen in the physical world!

To see this, let us perform a thought experiment. Imagine two airtight compartments, the first one empty (a vacuum) and the second one full of air. Open up some communication between them, a small hole for instance. The air in the second compartment will rush into the first one until air pressure is the same in both compartments. At this stage, anyone who witnessed the air in the first compartment spontaneously rushing back into the second one, and the second compartment emptying again, would not believe his eyes.

1. *Méthodes nouvelles*, 1:3.

There is a standard mathematical model for this physical situation. The gas is considered to be a collection of molecules moving in straight lines until they collide with the side of the compartment, or with another molecule. All collisions are assumed to be elastic, that is, no energy is lost.

A celebrated mathematical result, Poincaré's recurrence theorem, applies to this situation. It predicts that, in some future time, the system will come back to its initial situation. As a matter of fact, it will do so again at some still later time, and again and again, infinitely many times.

It is thus predicted that the first compartment, after being filled, will eventually empty itself completely into the second one. The air will then flow back, so that both compartments will contain about the same number of molecules, but eventually the first compartment will again empty itself into the second one, and so on to infinity. One could hardly imagine anything more contrary to everyday experience, or to the laws of thermodynamics.

If this really worked, it would provide humanity with a simple way of repairing a flat tire: just jack up the car, wait for the air to come back into the tire, and then mend the hole to keep the air in. In the same vein, if someone has put one lump of sugar too many in your coffee, just wait patiently till it crystallizes back again, at the bottom of your cup. For it is a mathematical certainty that the sugar which dissolved will

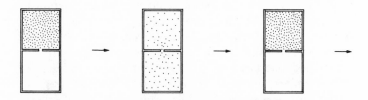

FIG. 2.2. Perpetual motion: according to Poincaré's paradox, the gas which has escaped from the upper compartment into the lower one will eventually stream up again through the hole and leave the lower compartment empty. The whole cycle will then begin again.

crystallize back, as surely as the air which left the tire will find its way back through the hole.

The solution of these paradoxes lies in the amount of time which is necessary to observe one of these cycles. If the air in the second compartment was very rarefied, containing only one molecule, this molecule would be found in each compartment half the time, and no one would find anything paradoxical in that. If there are two molecules, there will be more possibilities, four in all, only one of which corresponds to an empty first compartment, which means one will still see it frequently, but will have to wait longer for it. Now, for realistic quantities (one liter of air contains about 2.7×10^{22} molecules), the amount of time we will have to wait before seeing the first compartment empty again can be computed, and it turns out to be greater than the age of the sun. This explains why these mathematical predictions have no practical relevance.

This is the second side of Poincaré's criticism. On the one hand, he shows us models which are exact but incapable of prediction, and, on the other hand, models which predict the impossible with certainty. In this way, he paves the way for a new type of model, which will indicate what possibilities the future holds in store without predicting which one will be chosen. Such qualitative models are as different from quantitative ones as a drawing is from a computation.

Poincaré was the first to introduce qualitative methods into the theory of differential equations. In the particular case of celestial mechanics (what he calls the equations of dynamics), he has documented the global complexity of the motion by investigating individual trajectories of a special kind, and the situation in their immediate neighborhood. In this fashion, he has unearthed situations the complexity of which was totally unexpected, and has demonstrated that the motions described by Newton's equations are as a general rule extremely irregular. Under the apparent macroscopic regularity of the Keplerian orbits, Poincaré has shown us a wealth of microscopic moves, analogous to the case of a particle which our eye sees at rest and which a microscope reveals to be agitated by Brownian motion.

Let us try to retrace his steps. The starting point is the investigation of some individual trajectories of a special kind: the periodic ones. A trajectory is periodic if it closes back upon itself after a certain time T, which is called the period. In other words, a trajectory is T-periodic if the corresponding motion goes through precisely the same positions at intervals of T. The orbit of the earth, for instance, is periodic in the Keplerian approximation, with a period of one year. The true orbit, taking into account planetary perturbations, is not very likely to be periodic. If it is, the period must be very long, much greater than one year.

Let us hear what Poincaré has to say about periodic trajectories (or orbits; the two words have the same meaning). His flowery language has fallen into disuse in today's science. "The reason we find these periodic solutions so irreplaceable is that they are, so to speak, the only place where this unapproachable rampart has crumbled, and where we can try to penetrate the fortress."[2] Indeed, there are two advantages to them: they can be computed, and we can describe the situation around them.

Let me stress the first point, because it is a partial answer to a question we raised before. How can one write down explicitly a relation $x = f(t)$, when we want it to be valid for all times t, even very large ones? If f is known to be periodic, the task is much easier, because it is enough to write down the relation for all times between zero and T, that is, on a finite interval. The values of $f(t)$ for all t can be deduced easily. For instance, if I punch in 1000 on my hand calculator, and press the sine button, I get E for error. But if I really want to know the sine of 1000, I just divide 1000 by 2π, the period of the sine function, and take the sine of the remainder, which is 0.82687954. This is why the periodic trajectories in the three-body problem can, in principle, be computed explicitly. Note, however, that even if all periodic solutions of the three-body problem were known, which is far from being the case, they would not by themselves provide a complete solution of the three-body problem, since most solutions are nonperiodic.

2. *Méthodes nouvelles*, 1:82.

Once a periodic trajectory is found, the next step in the investigation consists in describing the situation in the immediate neighborhood. Let T be this periodic trajectory; it is a closed curve in the space. Let us cut it transversely by a vertical plane π, which will intersect T in a point which we call O (and in one or more other points we forget about). If now T′ is another trajectory, close to T, it will meet π at points $A_0, A_1, A_2, \ldots$, close to O. There will be infinitely many such points A_i, unless the trajectory T′ itself is periodic, which is not very likely. Poincaré's idea was to replace T′ by the sequence A_i of its points of impact with the plane π. The situation near T can now be described by a two-dimensional picture.

Just imagine that the plane π is a white sheet of paper, and that O is the point where the periodic trajectory T meets π. If we pick any other point A_0 near O in this plane, the trajectory going through A_0 in the space will stay near T, and will come back to hit π again near O. This defines a new point A_1 in the plane. After crossing π at A_1, the same trajectory, still following T in the space, will come back near O a second time, hitting

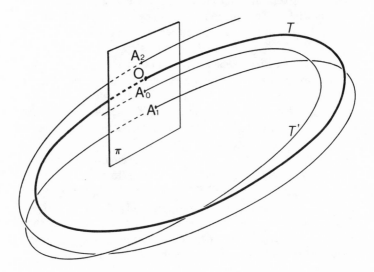

FIG. 2.3. The situation in the neighborhood of a closed trajectory

π at some point A_2. On our sheet of paper, which represents π, we will plot the sequence $A_0, A_1, A_2, \ldots$ of points of impact. It will enable us to visualize the trajectory T', which winds around T in the space as the sequence A_i winds around O in the plane. If, for instance, $A_0 = A_n$, i.e., the *nth* point coincides with the initial one, then the trajectory T' is periodic. It closes up after *n* turns, whereas T closed up in 1 turn.

With a personal computer this is an easy job to do, and the result can be fun. The reader may want to try his/her hand at it. Define the transformation $A_n \rightarrow A_{n+1}$ in the plane with Cartesian coordinates (x, y) by the following equations:

$$x_{n+1} = x_n \cos \alpha - (y_n - x_n^2) \sin \alpha,$$
$$y_{n+1} = x_n \sin \alpha + (y_n - x_n^2) \cos \alpha.$$

The angle α is a parameter which can be chosen freely. This simple example is due to Hénon,[3] and since the transformation $A_n \rightarrow A_{n+1}$ is given directly, there is no need to compute the portions of trajectories lying between A_n and A_{n+1} in the space. The computation is already done, and the result embodied in Hénon's formula.

Figures 2.4 and 2.5 describe a typical trajectory near a periodic orbit, represented by O. They have been drawn by taking $\alpha = 76°.11$ in Hénon's formulas. Various starting points A_0 have been chosen, so that in fact several trajectories are represented together on each of these pictures.

In Figure 2.4 notice the three central rings. They are associated with three different trajectories (that is, three different choices of A_0). Each of these rings seems like a continuous curve—but is not. The successive points A_i are spaced very regularly, and they fill out a curve, the image of which appears gradually as the number of points increases. The same number of points has been chosen in each case, and one can see how the inner ring fills out more quickly than the outer one.

As the initial point A_0 is chosen farther away from O, the resulting trajectory changes. The points A_i no longer fit on a nice smooth curve, but tend to fill out whole regions of the

3. "Numerical Study of Quadratic Area-preserving Mappings," *Quarterly Journal of Applied Mathematics* 27 (1969): 291–312.

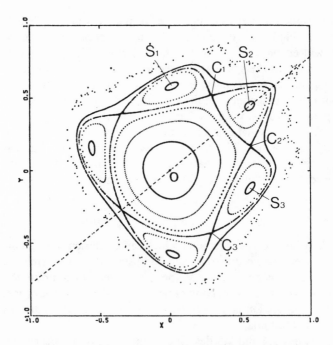

FIG. 2.4. Several trajectories of the Hénon system. Notice how
the inner trajectories seem regular (the successive points are
distributed on circles), whereas the outer trajectories seem chaotic
(the successive points seem to be randomly distributed over a whole
area). From M. V. Berry, "Regular and Irregular Motion," in
Topics in Nonlinear Dynamics, AIP Conference Proceedings, no. 46.
© 1978 American Institute of Physics.

plane. The halo which surrounds the picture on the outside is
constituted by a single trajectory A_i which apparently wanders
aimlessly. Between this one and the inner trajectories we find
an intermediate region, where weird things happen.

We first see five "islands," culminating at S_1, S_2, S_3, S_4, S_5,
separated by five "straits," C_1, C_2, C_3, C_4, C_5. In each of these
islands, the "level curves," which become more and more distinct as they get closer to the summit, are created by a single
trajectory (two trajectories for the two level curves the picture
shows). So each of the summits S repeats the structure we saw
around O, with 5 times the period. A trajectory starting from

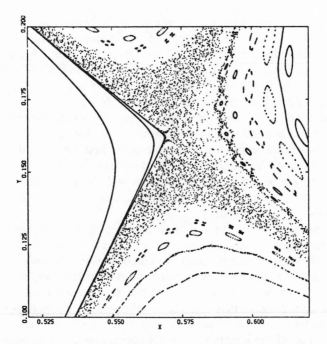

FIG. 2.5. An enlargement of the region around point C_2 of Fig. 2.4. The points on the picture belong to a single trajectory. Notice the islands enclosed within the chaotic sea. From Berry, "Regular and Irregular Motion."

a point A_0 close to S_1 will first hit a point A_1 close to S_2, then a point A_2 close to S_3, then a point A_3 close to S_4, then a point A_4 close to S_5, before coming back close to S_1 at a point A_5. This single trajectory is building five "level curves" simultaneously, one in each island. As a particular case, the trajectory coming out from S_1 goes through S_2, S_3, S_4, S_5, and then back to S_1. So it is periodic, and its period is about 5 times the period of T (going through O).

The straits C_1, C_2, C_3, C_4, C_5 also belong to a single periodic orbit. We have thus found, around the original trajectory T, an inner region where the behavior is regular, an outer region where the behavior is chaotic, and an intermediate region which contains two periodic orbits with 5 times the period of T.

There is much more to be said about the complexity of the situation. Let us look at Figure 2.5. It is a blowup of the point C_2 enlarged about 30 times. All the dots on the picture belong to a single trajectory. They are spread over a two-dimensional region, the chaotic sea. We now discern a finer structure, which was hidden at a larger scale. Curves which seemed to be well defined dissolve into a vague halo dotted with islands. Another blowup, on a still smaller scale, would show that each of these islands again reproduces the structure we first saw around O. Each island is a microcosm, a faithful reproduction of the whole: it will contain smaller islands, which will also reproduce the general structure, hence contain still smaller islands, and so on to the infinitely small.

We now see a hierarchical structure of great complexity emerging gradually. Several pictures come to mind. We may think of a sponge, structured by holes of all sizes. We may also recall those posters showing some character holding the same poster, where you see in reduction the same character holding the same poster, where you see. . . . Remember also what happens when you place two mirrors in front of each other, and you see all the images of yourself converging down to some point at infinity. Those of us who have been to the Soviet Union have probably brought back a wooden "matriochka" doll, which opens to reveal another doll, which opens to reveal a third doll, which opens. . . . In the same way, the two pictures we just saw contain their own reduced image. The structure is the same at all scales. There is no difference between the microscopic and the macroscopic.

This structure makes possible a continuous transition from the regular, predictable, motion of the central trajectories to chaotic, unpredictable, motion of the outer trajectories. They are all deterministic, since they arise from a differential equation, but an observer facing the pictures we just saw would be more likely to attribute the chaotic trajectories to some kind of random motion.

The global picture mixes order and chaos very intimately. A trajectory which appears to be regular at a certain scale might show itself to be very perturbed at a smaller scale. However, in the chaotic regions which the blowup has revealed, islands of order will be found, and so on, repeating the same structure to smaller and smaller scales. Order and chaos, regularity and unpredictability, are woven together like land and sea on the beach when the tide is drawing out, leaving behind a maze of puddles and wet sand, so that it is impossible to tell where the water ends and where dry land begins.

Even periodic trajectories, which should be the epitome of regularity, reflect the insidious way in which chaos sets in. In the preceding pictures we have seen many of them. First the original trajectory, going through O, then the two trajectories $S_1S_2S_3S_4S_5$ and $C_1C_2C_3C_4C_5$. Take the original period to be 1, so that the periods of the two other ones will be about 5. Since the islands reproduce the global structure, they too must contain periodic trajectories of higher period—5 times the period of $S_1S_2S_3S_4S_5$, that is, 25 times the original period. In this way, we see that the picture must contain periodic trajectories of higher and higher period, 1, 5, 25, 625, 3125, and so on. For the latter, we would already have to construct more than three thousand points to check that it is periodic. It is very unlikely that such a trajectory would be noticed at all: a casual observer would see it as just a chaotic trajectory like so many others.

Poincaré himself had nothing like our computing facilities. He was led to this kind of picture by theoretical considerations, based on qualitative methods, instead of numerical simulations. He classified periodic solutions into two types, which he called elliptic and hyperbolic, and showed that, provided that certain exceptional cases were excluded, the local situation around an elliptic periodic trajectory was given by the pictures we just saw. His conclusions also imply the existence of a sequence of periodic orbits of higher and higher period, each of which gives rise to islands and straits, thereby proving rigorously the results we derived from an inspection of the pictures.

These conclusions will not be found in the *Méthodes nouvelles de la mécanique céleste*. This is because Poincaré reached them much later. As a matter of fact, there was still something he couldn't prove. Everything hinged on a geometric theorem, which he knew how to prove in many special cases but not in

full generality. Toward the end of his life he resolved to publish
it, so that younger mathematicians might succeed where he
had failed. It became known as "Poincaré's last theorem" and
was finally proved in 1913 by a young American mathematician,
George D. Birkhoff.

What will be found in the *Méthodes nouvelles*, however, are
investigations concerning hyperbolic periodic trajectories. This
led Poincaré to consider some nonperiodic trajectories of a
special type, which he called "doubly asymptotic" and which
nowadays are known as homoclinic trajectories. I am not going
into this, referring the interested reader to Appendix 1 for a
study of homoclinic trajectories and for some nice pictures.
Here is what Poincaré had to say about them: "One is struck
by the complexity of this picture, which I do not even attempt
to draw. Nothing can give us a better idea of the complication
of the three-body problem, and more generally of all the problems in dynamics where there is no uniform integral and the
Bohlin series diverge."[4]

Deterministic but Random

What do we have to offer now? The ancient temple has been
demolished; what shall we build in its place? We were used to
the Keplerian orbits, plane, elliptical, and periodic, completely
predictable. Like an ancient icon that has captured the people's
faith, it confirmed us in the belief that the earth orbits the sun,
today, tomorrow, and forever. The icon has been burned, the
Keplerian ellipse has dissolved into a halo, and no one knows
how long the earth will orbit the sun. What new image can we
hang up instead of the old one?

The first image that comes to mind is die throwing. It is
certainly a popular one, since Julius Caesar first used it while
crossing the Rubicon, and it captures many important features
of the motions we have just described. It is deterministic but
random, as they are. More precisely, the physical laws regulating the die's motion are purely deterministic, but the outcome is seen as random.

4. *Méthodes nouvelles*, 3:389.

What could be more deterministic than die throwing? This little cube, once it leaves my hand, is subject only to gravitational pull and air resistance. It rebounds several times on a surface which is made as plane and elastic as humanly possible, and stops after dissipating its energy in shocks and friction. The mechanics of the motion have been much studied and are well understood. In principle, at least, once the initial impulse is given, everything can be computed explicitly.

On the other hand, die throwing has always been considered the symbol of randomness. In Romance languages, the word for "random" derives from the Latin word *alea* ("dice"). Randomness itself is very difficult to define. It is mostly seen as an empirical fact, the typical example of which used to be die throwing, and may nowadays be found in quantum mechanics. The main feature of randomness is some degree of independence from the initial conditions. Even though the initial conditions are known, the outcome cannot be fully predicted, and the only way to know it is to actually carry out the experiment. Better still, if one performs the same experiment twice with the same initial conditions, one may get two different outcomes. The basic idea of playing dice is that you cannot (or should not) control the result of a throw by the way you hold the dice. Unpredictability is the name of the game.

So throwing a die may be seen as both a deterministic and a random event. Celestial mechanics also has this double character. The laws of motion are purely deterministic. But certain trajectories are so irregular, the halo in Figure 2.4, for instance, or the cloud in Figure 2.5, that they seem to arise from some kind of random motion. They certainly become unpredictable very quickly.

But this analogy is not a very good one: there are basic differences between die throwing and classical mechanics. It is easily seen that the two aspects of die throwing appear at different scales: it is deterministic on a small scale, and random on a large scale. Two initial conditions which appear to be the same on the macroscopic scale, the one the players perceive, will in fact be very different on the microscopic scale, which determines the motion, and hence will result in two different outcomes. On the other hand, as we pointed out earlier, celestial mechanics is the same on all scales. The structure pictures shown in Figures 2.4 and 2.5 must be repeated on finer and finer scales. Microscopic and macroscopic events are essentially the same.

We need a better image, which will capture these features of celestial mechanics, in addition to other ones we have not yet mentioned. Such an image exists. Mathematicians did not stumble upon it by chance or conjure it from literary reminiscences. After Poincaré, three more generations of mathematicians worked on dynamical systems, discovered the "baker's transformation," and showed that it provided a precise picture of what was going on in celestial mechanics. The baker's transformation is also called the "Bernoulli shift." Bernoulli was a Swiss mathematician of the seventeenth century, who finds himself associated with modern American (Birkhoff, Smale, Ornstein) and Russian mathematicians (Kolmogorov, Arnold, Sinai) in providing the twentieth century with its new image of dynamics.

Watch the baker at work. He takes some dough, works it into a thin sheet with a pastry roll, than folds the dough over itself once or twice, and rolls it down again. We shall make this process mathematically precise. At each stage, the thickness of the sheet will be halved and its length correspondingly doubled, its width remaining the same. The sheet shall then be cut in two, instead of folded over, and one of the half-sheets (always the same one) neatly put on top of the other. The process then starts again, with a sheet the same size as before.

The two first stages are shown on Figure 2.6. The top square represents the sheet of dough, seen from the side. It is then worked down to half its original height and double its length. The right half of the rectangle is then cut away and put back on top of the left half. The process is repeated, and the resulting (bottom) square has four layers of dough. All this becomes much more striking if one draws a cat's head on the initial square, and follows its successive transformations. The poor animal is called "Arnold's cat."

At the second stage, Arnold's cat has already been turned into mincemeat. Notice how the layers alternate: the first and

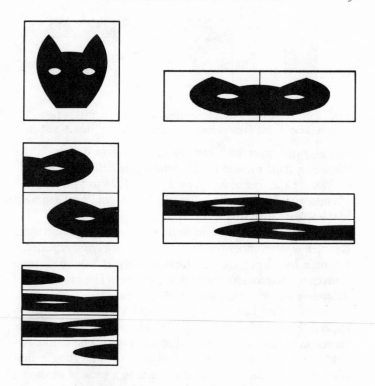

FIG. 2.6. Arnold's cat

third layers were part of the same layer at the preceding stage, and they are now separated by the second layer. Note also the possible discontinuities: points A and B are pretty close, but after two stages, their transforms A'' and B'' are far apart.

The situation becomes more complicated as the baker iterates the transformation. After 10 stages, there will be 1024 layers, and more than one million after 20. All these layers have been shuffled like a pack of cards. Arnold's cat has melted into the square, gradually disappearing from sight like the Cheshire cat in Wonderland.

But it can reappear, as easily as the Cheshire cat did. All the baker has to do is to pull up his dough instead of flattening it, and when the dough reaches twice its original height, cut it horizontally, and lay the top half beside the bottom. Alter-

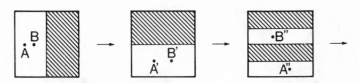

FIG. 2.7. The baker's transformation. The two first steps have been indicated, with the successive transforms of points *A* and *B*.

natively, the baker may just lay his square of dough on the side, then flatten it and cut it vertically as before. Doing this divides the number of layers by 2, from 1024 to 512, say, and after doing it 10 times the baker will find Arnold's cat grinning in the square it started from.

This is a typically deterministic process. The present state completely determines the future ones by applying repeatedly a simple law. Knowing the situation fully at any given time enables us to reconstruct any past situation. The past and the future are entirely contained in the present.

On the other hand, this process quickly leads to chaos—apparently to total disorder. Any small piece of the square is spread more and more evenly throughout the whole, like Arnold's cat, until there is an equal likelihood of meeting it anywhere in the square. Repeatedly shuffling a pack of cards is a process of this kind, and is also an accepted symbol for randomness. The underlying idea is that, if the job is done properly, the likelihood of finding the ace of spades anywhere in the pack is the same.

A full-fledged mathematical theory has been developed to study the way small pieces of the square are spread around by the baker's transformation, or by similar ones. It can be characterized by a single number, called the entropy of the transformation. We are not going to delve into this rather technical subject, which is open to many misunderstandings and undue generalizations. Instead, we shall concentrate on individual trajectories, the successive transforms of a single point. They will be quite enough to indicate how the process leads to chaos.

In Figures 2.8 and 2.9 two different trajectories are depicted. The initial point is labeled *0*, its transform *1*. The transform

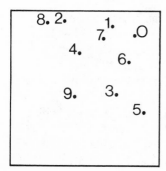 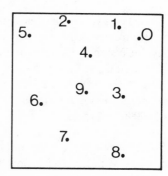

FIG. 2.8. The initial point *0* has coordinates $x = y = 0.840675437 \ldots$

FIG. 2.9. The initial point *0* has coordinates $x = y = 0.846704216 \ldots$

of *1* is the point *2*, the transform of *2* is the point *3*, and so forth. We shall also refer to the point *1* as the first iterate of *0*, to the point *2* as the second iterate of *0*, and so on. The infinite sequence of all such iterates is the trajectory of the point *0*. In Figures 2.8 and 2.9 we have shown the first nine iterates only. Notice that, although the initial points have been chosen rather close to each other, the trajectories separate very quickly. Notice also that the trajectories tend to spread out evenly into the square. Pushing the experiment further, computing hundreds or thousands of iterates instead of nine would lead to the same conclusions.

Plotting trajectories in the square is a very inefficient method. It becomes useless if large numbers of iterates must be computed, or, if one wants to study complete trajectories, the full infinite sequence of iterates. A very clever method has been developed for this purpose.

Let us change our modus operandi slightly. At each stage, the baker will make 10 layers instead of 2. He will work down the square to one-tenth its initial height and 10 times its length, cut it into 10 strips of equal length, and put them on top of one another to build a 10-layered square.

Each point in the original square can be conveniently defined by two numbers: its distance from the left side and its height. Say, for instance, that the sides of the square have length unity. The point C_0 in Figure 2.10 is exactly at mid-height, and one-third of the way across from the left side: its position in the square is then given by the two numbers 0.333333 . . . and 0.500000. More generally, any two decimal numbers, with integer part zero, will define a point in the square. Let us pick them at random. I whip out my hand calculator, push the RANDOM button twice, and get 727 and 756: write 0.727756 for the first number. Proceeding similarly for the second number, I get 0.578675. The two numbers 0.727756 and 0.578675 define a point D_0, which is shown in Figure 2.10 with an accuracy limited to the second decimal place.

The (10-layered) baker's transformation can now be written very simply. Remove the first decimal of the first number and make it the first decimal of the second one. For instance, the transform C_1 of C_0 is defined as follows:

$$C_0 \quad \text{by} \quad 0.333333 \ldots \text{ and } 0.500000,$$
$$C_1 \quad \text{by} \quad 0.333333 \ldots \text{ and } 0.350000.$$

The transform D_1 of D_0 is defined as follows:

$$D_0 \quad \text{by} \quad 0.727756 \text{ and } 0.578675,$$
$$D_1 \quad \text{by} \quad 0.27756 \text{ and } 0.7578675.$$

In the same way, we find the further iterates of C_0 and D_0. They are defined as follows:

$$C_2 \quad \text{by} \quad 0.333333 \text{ and } 0.335000,$$
$$C_3 \quad \text{by} \quad 0.333333 \text{ and } 0.333500,$$
$$C_4 \quad \text{by} \quad 0.333333 \text{ and } 0.333350,$$

and

$$D_2 \quad \text{by} \quad 0.7756 \text{ and } 0.27578675,$$
$$D_3 \quad \text{by} \quad 0.756 \text{ and } 0.727578675,$$
$$D_4 \quad \text{by} \quad 0.56 \text{ and } 0.7727578675.$$

We now see why the baker's transformation is also called the Bernoulli shift: It consists in shifting the decimal point one

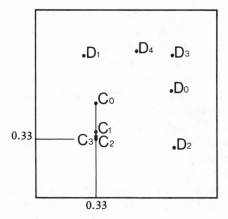

FIG. 2.10. The decimal baker's transformation, also called the Bernoulli shift

space to the right in the first number, which defines the horizontal position, and one space to the left in the second number, which defines the height.

Now that we know how to generate individual trajectories, we can start investigating them. Imagine, for instance, that the second number only is available for observation. In other words, we can determine the height of any point in the square, but not its horizontal position (distance from the left side). This would be the case for an observer located to the side of the square, and in the same plane. The (two-dimensional) square appears to him as a (one-dimensional) line, and he is aware of the vertical component of the motion only. Horizontal displacements of points in the square are hidden from his view. He loses one of the two dimensions of the baker's transformation.

Let us see what such an observer makes of the two preceding examples. In the first case, a trajectory initiated at C_0, he sees a point located at 0.500000 move to 0.350000, then to 0.335000, then to 0.333500, and so forth. The successive iterates move ever closer to the point 0.333333 . . . = $\frac{1}{3}$, without actually reaching it. If the observer was already in position before $t = 0$, he will have seen the point at 0.000000 . . . when $t = -1$, $t = -2$, and all preceding times. As far as he can tell, the history of this motion is as follows: a point, which was at rest during all the past, suddenly starts moving when $t = 0$. It first moves to $\frac{1}{2}$, then inches toward $\frac{1}{3}$, getting ever closer to that final position without actually reaching it.

In the case of the second trajectory, initiated at D_0, the observation jumps back and forth along the segment in an unpredictable fashion, from 0.57 . . . to 0.75 . . . to 0.27 . . . to 0.72 . . . to 0.77, and so on. There seems to be no pattern to these jumps, and the observer has no way of predicting where in the segment the next iterate will lie. As far as he can tell, this is random motion.

Let us now put a lamp where our observer was, so that each point in the trajectory we are observing will project a shadow. We collect these shadows on a screen, where they will only move up and down along a line. The second dimension is still lost, since the lamp lights the square from the side. We now let our observer watch the screen. The result is the same as before, but now we have a modern version of Plato's famous myth.[5] Prisoners are chained in a cave, their backs to the entrance, facing the end wall, while their captors are passing by the entrance, carrying puppets and various objects which the sunlight projects as shadows on the end wall. His theme is that if the prisoners have been in that situation for generations, they will forget what the outside world was like, or that it ever existed, and will take the shadows to be the real things. Plato sees mankind in this situation, the true reality belonging to the world of ideas, and projecting down imperfectly into the world of appearances. Our projection lamp is a modern version of Plato's myth. It changes determinism into randomness, as substance was changed into mere appearances in Plato's cave.

Why do we mention randomness in the context of the Bernoulli shift? Let us first realize how powerless the observer becomes to understand the process when the horizontal position is withheld. At any given time, $t = 0$, say, he knows the complete past history: he has at his fingertips all the past observations, as far back as he cares to go. Indeed, since he knows

5. *Republic* 7.514A–521B.

the position at $t = 0$, 0.578675 . . . , for instance, he obtains the position at $t = -1$ by shifting the decimal point one space to the right, 0.78675 . . . , the position at $t = -2$ by shifting it two spaces, 0.8675 . . . , and so forth. If all numbers in the decimal expansion, symbolized by points, are known, we can go back in time as far as we want. The past is an open book.

On the other hand, the future is completely closed. We are at a total loss to predict what is going to happen at $t = 1$. There are 10 possibilities, from 0.0578675 . . . to 0.9578675, through 0.1578675. It so happens that, in the trajectory under consideration, beginning at D_0, the actual position will be 0.7578675 . . . , but there is nothing in the past history which could help us to predict this. To be sure, we know what all the decimal digits are going to be, except for the most important—the first one. This means that we cannot tell even very roughly where the next observation will be, in what region it will fall. This will only get worse as one tries to look deeper into the future. If we know the position at time $t = 0$, we have no idea of the n first decimal digits of the position at time $t = n$, which leaves us with 10^n possibilities spread evenly along the segment.

Let us try another thought experiment. Our observer, still looking at the shadows moving up and down on the screen, has now acquired a small monkey, which plays at his side. The animal finds a die on the ground and starts to throw it. His master is happy to find some relief from boredom, and jots down the numbers as they appear, 436345, say. This would certainly qualify as a random sequence of numbers. Suppose, now, that the observations on the screen seem to conform to that particular random sequence: say, for instance, that the first decimal place, the unpredictable sequence, is successively a 4, then a 3, then a 6, then a 3, then a 4, then a 5. Our observer would have sufficient reason to conclude that he is watching a random process, as random at least as throwing a die.

This particular sequence of observations is perfectly possible. Just start from a point on the square defined by 0.436245 . . . and 0.000000. . . . The sequence of observations from $t = 0$ will then be

for $t = 0,$ 0.000000 . . . ,
for $t = 1,$ 0.400000 . . . (4 appears),
for $t = 2,$ 0.340000 . . . (3 appears),
for $t = 3,$ 0.634000 . . . (6 appears),

and so on. Clearly any particular sequence of any length can be produced in this way.

This is really what is going on when one presses the RANDOM button of a hand calculator. Any kind of computer must be a purely deterministic machine. How can it produce random sequences of numbers? That, of course, touches upon the question of what is meant by randomness. What the computer does is to produce, by purely deterministic processes, sequences of numbers which mimic the kind of sequences that can be obtained by throwing a die. Such sequences will be called random, just as the distribution of points in Figure 2.5 seemed to be random, or the vertical displacements of a point in the Bernoulli shift were unpredictable. There are many ways for a computer to produce such sequences. One of the simplest is truncated multiplication. Take a 6-digit number, and produce a new one by multiplying by itself, and erasing from the result (which has 11 or 12 digits) the first 2 or 3 and the last 3 digits. We can repeat the process, and it is clear that after a few iterations the 6-digit number we obtain has little relation to the original one. For this reason we call it random. In this way, randomness is achieved as the result of a computation.

Let us go over this again. The Bernoulli shift is purely deterministic. However, by looking at it in a certain way, incomplete to be sure but accurate (there is no error or uncertainty in the observations), we find that the observations follow a random pattern. The next question is whether we are just playing with words or whether we are describing some physical reality. Is the baker's transformation anything other than a perverse way for mathematicians to have fun? Can we see this kind of randomness in more realistic experiments?

We can. Thanks to the mathematicians we mentioned, particularly Birkhoff and Smale, we know today that in celestial mechanics certain kinds of motions can be identified with Bernoulli shifts. Figure 2.5, for instance, shows the successive points of intersection of a trajectory with some transversal plane. It looks very much like what Figures 2.8 and 2.9 would become if we iterated the baker's transformation a sufficient number of times. This is not only a similarity: it can be proved that the baker's transformation and the kind of motion we observe in Figure 2.5 really are one and the same. Everything we observed about the Bernoulli shift must therefore have its counterpart in celestial mechanics.

We should therefore be able to show some randomness in celestial mechanics, even though it operates under the strictly deterministic laws of gravitation. One last loophole is left to the believers in classical determinism: will any kind of randomness perhaps be confined to small-scale phenomena, leaving our macroscopic world more or less completely predictable? That this should not be so is apparent from our previous discussion, where we found the underlying structure to mirror itself at the various scales, and where we concluded that there should be no difference between microscopic and macroscopic phenomena.

Here is a famous example. Imagine two stars of equal mass, rotating around their common center of gravity. Newton's law asserts that this is possible, and that the orbit of the two stars will be a circle, along which they travel with equal speed, being exactly opposite to each other at all times. Let P be the plane containing the circular orbit and D the axis of symmetry, which is perpendicular to P.

A third body, with very small mass—a comet, for instance—moves along D. It again follows from the laws of celestial mechanics that this is possible: if the motion starts on D, and the initial velocity is along D, then the comet will stay on D forever. It will move back and forth along D, crossing the plane P in both directions along its way.

It so happens that near one of the stars there is a planet inhabited by intelligent beings, and that the appearance of the comet has extraordinary religious significance for them. They can see the comet only when it gets close enough to the stars—that is, when it crosses the plane P. They have kept very precise records of the comet's appearances, and they know that the last appearances of the comet were 17, 35, 143, and 305 stellar

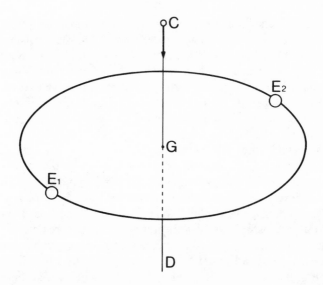

FIG. 2.11. The comet is observed only when it crosses the plane
where the orbits of the two stars in the system lie.

years ago (a stellar year is the time it takes the twin stars to
run once through their orbit). We can even stipulate that they
have recorded every appearance of the comet since creation.
With this very complete list of observations, they come to the
Court Astronomer and ask him: "When will we see the comet
again?"

The only true answer this unfortunate gentleman can give—
though not the safest one—is, "I couldn't possibly know." The
comet may show up today, one year from now, in ten years,
in a thousand years, or never again. All these various possibilities fit in with the available information! No computation
the astronomer could make will enable him to predict any one
of these possibilities, or even to state that some are more likely
than others. The fact is that, as Sitnikov has shown,[6] in this

6. "Existence of Oscillating Motions for the Three-Body Problem," *Dok-
lady Akademii Nauk USSR* 133 (1960): 303–6. See also V. Alekseev, "Qua-
sirandom Dynamical Systems. I, II, III," *Mathematics of the USSR (Sbornik)*
5 (1968): 73–128; 6 (1968): 505–60; 7 (1969): 1–43.

particular case the laws of celestial mechanics make any sequence of observations possible! The sequence we just indicated,

$$\ldots -305, \ -230, \ -143, \ -35, \ -17,$$

can be continued in any fashion at all. It can be perfectly regular, as in

$$1, 2, 3, 4, 5, 6, 7, \ldots$$

or

$$10, 100, 1000, 10,000, \ldots,$$

or totally irregular, as in

$$72, 757, 9431, 9432, 102437, \ldots.$$

Any of these sequences can be achieved physically. Whatever the record of past observations was, and whatever sequence of numbers is chosen, there is always a trajectory of the comet, obeying Newton's law, which will fit the past record and which will cross the plane P at the indicated times in the future. In other words, future observations are totally independent of past observations. Knowing all the past appearances is no help in trying to predict the next one, any more than knowing ten consecutive draws at the roulette wheel will help to guess the eleventh. This independence of the future from the past is randomness, as opposed to determinism.

Imagine what philosophy and science would be like in such a world. We look at the night sky, see the moon and the planets moving more or less regularly against a fixed background, and conclude that there is a basic regularity in nature. For these people, looking at their night sky and waiting for the comet, the first basic experience would be the unpredictability of nature. They would readily believe that God plays dice, a suggestion which Einstein, in his famous controversy with Bohr, indignantly rejected.

We recognize in all this the baker's transformation. These sequences of past observations which can be continued into the future by any sequence we care to use are already familiar to us. We saw them as moving shadows on a screen, while a small monkey was playing by our side. They now have a much bigger stage, but the underlying mechanism is still the same.

Randomness appears because the available information, though accurate, is incomplete. Part of the information is withheld from us. The spectator at our slide show does not know the horizontal position of the point, though he follows its vertical motion on the screen. The astronomer in the twin star system does not know how fast the comet moves when it crosses the plane P. If he did, he would have no problem computing the full trajectory of the comet, checking it against past observations, and predicting the next appearances. But he only knows the dates when these crossings occurred, and, strangely enough, not knowing the speed forbids him to make any prediction at all. It is all the more remarkable because the question that is raised, namely, the date of the comet's next appearance, makes no reference to the missing information, the speed at which it crosses the plane P. One would have thought that the answer relied more on past information concerning the dates of appearance, which turn out to be fully known but useless, than on any information about the speed.

So, if determinism means that the past determines the future, it can only be a property of reality as a whole, of the total cosmos. As soon as one isolates, from this global reality, a sequence of observations to be described and analyzed, one runs the risk of finding only randomness in that particular projection of the deterministic whole. Not that we have any choice. Global reality, the cosmos taken as a whole, from the most minute elementary particle to the expanding universe, is out of our reach. Science can only isolate subsystems for study, and set up experimental screens on which to project this inaccessible whole. Even if reality is deterministic, it may well happen that what we observe in this way is unpredictability and randomness.

From a strictly scientific point of view, there is only one thing we can apply the laws of physics to, and that is the universe. There is no physical subsystem which we could isolate from the influence of the rest of the cosmos. An electron, located at the boundary of the known universe, will still make its presence felt on the earth, by its gravitational pull and electromagnetic field. Quantum mechanics even tells us that its wave function does not vanish, so that at any time this electron has a certain (extremely small, but not zero) probability of being found on earth. Clearly, such interactions are extremely small, but they add up, and to neglect them, as one must do to keep the problem within reasonable bounds, may lead to wrong conclusions. We shall see an example of this in the next section.

Let us just insist that the only object the laws of physics can be applied to with total confidence is the universe as a whole. It is the only physical system which contains all the information necessary for applying the laws of physics with perfect accuracy. In theory, a complete and detailed knowledge of the state of the universe today would be necessary to make any predictions about the future, if full scientific rigor is to be observed. In practice, no such knowledge can be attained or even remotely approached. What we do is to carve out subsystems, to which we apply the laws of physics as if they were isolated. Throughout this book, for instance, we have studied the solar system as if there was nothing else in the universe, no other stars or galaxies to perturb the motion of the planets. By doing this, we are throwing away some information, which might or might not be relevant. We let go of the total deterministic system, which we cannot handle, and confine our observations to narrow subsystems, trying to interpret them independently of the rest. We go into Plato's cave, turn our backs to the sunlight, and watch the shadows on the wall.

These subsystems may exhibit randomness, even though the total system is deterministic. This is the lesson the Bernoulli shift teaches us. Like the queen of England, determinism reigns but does not govern. Its power nominally extends over vast territories, where local rulers are in fact independent, and even turn against it.

The beautiful regularity of Kepler's law, and the ensuing predictability of the solar system on the human scale, is an accident in celestial mechanics. On another time—or length—scale, the motion of planets would become irregular and unpredictable. The picture to carry in our minds is Arnold's cat, as a symbol of the baker's transformation, and a reminder of the fact that a purely deterministic law may materialize in a totally random sequence of observations if part of the information is withheld, as it must be in any practical situation.

UNSTABLE BUT STABLE

Once upon a time there was a meteorologist named Lorenz. He began working—and still works—at a time when digital computers were changing the way scientists were doing research. Machines could compress a lifetime of computations into a few minutes, and could henceforth be used to perform "numerical simulation," that is, to test a mathematical model by its performance in various situations. A model for weather forecasting, for instance, would be tested by verifying that its large-scale behavior is similar to the real-world year-round situation (there should be a monsoon in Southeast Asia, and an anticyclone near the Azores) and that it approximates reasonably well the finer structure, that is, the way perturbations propagate (where are tropical storms generated, and where do they go?). This kind of testing is far beyond the computational powers of an individual.

At the time (we are talking about the 1950s) there was no more confidence than there is today in weather forecasting. Meteorologists themselves were not sanguine about their predictions. Strangely enough, there was no problem with the mathematical model: the underlying physics was well understood, and the equations were safely written down. They were complicated, to be sure, even very complicated, but they were known, and they should have been helpful in predicting the weather with some degree of accuracy. Unfortunately, this was not the case: forecasting the weather for tomorrow went smoothly enough, but as far as the weather for next week was concerned, mathematical models and last-generation computers did not perform much better than an educated guess.

Lorenz tried to get to the heart of the matter by simplifying the equations. He simplified them so much that one may doubt that the final product had anything to do with weather forecasting, but the result was a system of three differential equations with three unknowns (x, y, z), depending on three constants (a, b, c):

$$\frac{dx}{dt} = -ax + ay,$$

$$\frac{dy}{dt} = bx - y - xz,$$

$$\frac{dz}{dt} = -cz + xy.$$

This may not mean much to the layman, but you can take it from me that differential equations do not come much simpler. The only complication arises from the rectangle terms, where one multiplies two variables, xz in the second equation and xy in the third. If one strikes them out, the remaining terms contain only one variable at a time, and the resulting equations can be solved explicitly by elementary methods. In other words, Lorenz's system is the simplest one which cannot be solved on sight.

It turns out that it cannot be solved at all. It is another case when there is no "complete solution," that is, no way to express the variables x, y, and z as explicit functions of time and initial positions. Instead, one can try a numerical simulation: the computer is fed the initial positions (x_0, y_0, z_0), and then computes the following positions: (x_1, y_1, z_1) at the time $t = t_1$, (x_2, y_2, z_2) at $t = t_2$, and so on.

This is what Lorenz did. He tried several different simulations starting from various initial positions and lasting several hours. At one point, after a particularly long computer run, he wanted a rerun of the final phase. He did the obvious thing: he entered into the computer as initial data the positions at the time from which he wanted the rerun, as he read them from the printout.

As he tells the story, when he came back an hour later, time enough for the computer to simulate about two months of weather, the solution he found waiting for him had nothing in common with the solution the computer had given on the earlier run. His first reaction was to blame some technical malfunction—nothing unusual in those days—but it became clear quickly enough that these two different solutions did not in fact come from the same initial data. The computer used six decimal digits in its calculations but printed only three, so that the new initial conditions were equal to the old ones, plus small perturbations. These perturbations grew exponentially, doubling every four days, so that after two months of simulated weather each solution was going its own way. His immediate conclusion was that if the true equations regulating the atmosphere behaved in the same way as this simpleminded model, long-range weather forecasting was impossible.

Lorenz's equations exhibit instability with respect to initial data. The slightest shift in the initial positions becomes amplified during the motion, and the resulting trajectory may be wildly different from the expected one. If we now recall how Lorenz got his equations in the first place, we understand why weather forecasting is so difficult. It is because the equations of meteorology have the same instability property which Lorenz's equations inherit. The most minute error in the observations, the slightest shift in the initial data, may result in a completely different picture. It is estimated that small perturbations are multiplied by 4 every week, or by 300 every month. This is what Lorenz calls the "butterfly effect." A butterfly's capricious flight may result in a tropical storm, not tomorrow, but one or two years down the road. This is why long-range weather forecasting is so difficult: everything, absolutely everything, must be taken into account. No perturbation can be deemed too small to have any influence.

We now know many mechanical as well as physical systems which have the same kind of instability, that is, which amplify initial shifts during the motion. This indicates another way by which deterministic systems may be perceived to be random. If the experimenter reproduces exactly the same initial position, he will observe exactly the same trajectory: this is what it means for a system to be deterministic. But in practice, the initial conditions can never be reproduced exactly. There has to be some discrepancy, be it ever so small. This discrepancy, which passed unnoticed at the beginning, will be amplified with time, resulting in the long run in a completely different situation. The system thus exhibits some degree of randomness: the (seemingly) same initial conditions may lead to significantly different evolutions. Experiments cannot be reproduced, since absolute accuracy cannot be achieved. This is precisely what happens when one is playing dice. If a die is thrown twice in precisely the same way, the same number will turn up twice. The problem is that no one can throw a die twice in precisely the same way, and this is why dice is a game of chance and not a skill. The same idea was expressed by Heraclitus in 500 B.C.: "There is no stepping twice in the same river, or touching twice some perishable substance in the same state."

More modern masters, like Maxwell or Poincaré, knew such systems too, and had drawn their own conclusions. Let us quote Maxwell (1873), for instance:

It is a metaphysical doctrine that from the same antedents follow the same consequents. No one can gainsay this. But it is not of much use in a world like this, in which the same antecedents never again concur, and nothing ever happens twice. . . . The physical axiom which has a somewhat similar aspect is that "from like antecedents follow like consequents." But here we have passed from sameness to likeness, from absolute accuracy to a more or less rough approximation. There are certain classes of phenomena . . . in which a small error in the data only introduces a small error in the result. . . . The course of events in these cases is stable. There are other classes of phenomena which are more complicated, and in which cases of instability may occur, the number of such cases increasing, in an extremely rapid manner, as the number of variables increases.[7]

7. Quoted by M. Berry, "Regular and Irregular Motion," in *Topics in Nonlinear Dynamics*, American Institute of Physics Conference Proceedings, no. 46, ed. Siele Jorna (New York: American Institute of Physics, 1978), pp. 111–12.

This should lead us to have a closer look at the way we apply the laws of physics. We said just a moment ago that, in principle, these laws can only be applied to the universe as a whole. In practice, they are applied to subsystems which we somehow isolate, either in our minds or in the laboratory. We then assert that the influence of the rest of the universe on the subsystem is negligible. For instance, when computing planetary orbits, we do not take into account the (presumably small) perturbations due to nearby stars or galaxies. This may lead to unsuspected pitfalls when the underlying system is unstable.

Let us say, for instance, that we are watching a game of pool, and one of the players is setting up his next shot. It does not enter his mind that he should take into account the fact that the gravitational fields of the spectators, including mine, will cause the balls to deviate slightly from their trajectory. Nor do I think that I may alter the outcome, and make him miss his shot, by the simple device of changing my place in the room. As a matter of fact, both of us are right, but it is a close call. The actual computation shows that the gravitational field of a spectator's body at the edge of a pool table does not significantly alter the trajectory of a cannoning ball if there are two impacts only. But it also shows that after nine impacts or more, the perturbation has become important! In other words, if a player wanted to cannon nine balls with a single one, he would have to take into account the position of the spectators in the room!

Now remember that a gas is nothing but a three-dimensional pool table with incredibly many billiard balls, all cannoning gaily into one another. The same calculation as before shows that an electron located at the edge of the known universe, say 10^{10} light-years away, makes its influence felt from the fifty-sixth impact on.[8] All this, of course, within the deterministic framework of Newtonian physics, without even appealing to the uncertainty principle of quantum mechanics.

The upshot is that, when systems are so unstable, there is no point in computing individual trajectories. One could at-

8. Berry, "Regular and Irregular Motion," p. 95.

tempt a numerical simulation for a billard table with three or ten balls (very far from the 6×10^{23}—Avogadro's number—which can be found in realistic quantities of gas), enter initial positions and velocities into the computer, and derive the individual trajectories. The result would very quickly lose any practical significance, first, because the computer makes round-off errors: it works with 12 of 24 digits and cuts out supplementary digits which appear after each multiplication or division. These errors are amplified very quickly, as in Lorenz's problem, and alter the final result. In addition, the real system which our computer is supposed to model is never isolated. Many perturbations affect it—the motion of an electron on Sirius, the experimenter's presence in the room—which the mathematical model does not take into account. As we have just seen, these perturbations quickly become significant, so that the predicted trajectory, even if there were no built-in computational error, would still be very far from the original one.

In the preceding section we saw how a deterministic system may exhibit random behavior if part of the information is hidden from the observer. The situation here is somewhat different. All the information is available: the only problem is that there is too much of it! Positions and velocities can be measured with as many decimal digits as one wishes, but not with the infinitely many decimal digits which total accuracy would require. There will always be a slight discrepancy between measurements and the "true" data. This discrepancy may grow quickly, resulting in important differences between the predictions and the actual evolution. The system is deterministic, but unpredictable in the long run.

This die is in our hands, and so are the differential equations which govern its motion. We should be able to throw it in precisely the right way to show a six. The whole problem is that it is an unstable system: it would require superhuman accuracy on the throw to guarantee the outcome.

So this is another side to the same fact which we have already met in celestial mechanics: computation may be powerless. Quantitative methods break down, and we again resort to qualitative methods. The question now is this: if we cannot predict individual trajectories, what is there left to study? What can science say about an unpredictable system?

For die throwing the answer has been known for a long time. One should forget about individual throws, and consider instead the set of all possible throws. It can then be asserted that there are six possible outcomes, and all are equally likely. We attribute to each of them a probability $\frac{1}{6}$, and this is the beginning of probability theory.

Similar results have been obtained since 1960 for more general unstable systems, such as the Lorenz equations. In this context we have to quote the names of Anosov and Ruelle, in addition to Smale and Sinai, whom we have already encountered. These mathematicians, and many others, have created a new field of investigation which we will now try to describe.

The main problem in understanding the Lorenz equations of meteorology, for instance, is to gain some understanding of the long-term behavior of the various trajectories. In die throwing this is very easy: the die will eventually come to rest on one of its six sides, corresponding to six possible outcomes. For the Lorenz equations, the situation is much more complicated, since the motion goes on indefinitely. The system does not come to rest as time elapses, but the motion does become simpler. The system, which originally was free to move in three-dimensional space, is eventually observed to confine itself within a narrow region, a kind of strip, very complicated but very thin, which is called a "strange attractor." In other words, the motion eventually takes place within the strange attractor. Motions on the strange attractor are "final motions," or "outcomes," and a statistical model can be adapted to them, in much the same way as the outcomes of die throwing can be described statistically. The analysis is much more difficult, of course, and is essentially due to Ruelle.

We cannot go into these probabilistic aspects, but we will investigate another very interesting question, namely, the shape of the strange attractor. It is neither a volume nor a surface, but is somewhere in between. This is hard to understand and even harder to draw, but a famous picture due to Smale, the "horseshoe," will help us.

Let us go once more into the bakery. The baker is still at work kneading the dough, but this time he is putting so much energy into his work that he is actually compressing the dough. As before, we represent the three-dimensional block of dough by its two-dimensional section, the area of which shrinks at each step. The initial square is rolled down and folded over (this time without cutting). We get a kind of horseshoe, which we have no problem drawing into the initial square, since the total area has shrunk. As Figure 2.12 shows, this amounts to sending the initial square into itself. Areas are shrunk in the operation, whereas in the Bernoulli shift they were not.

Now we can have our fun. What happens to the horseshoe itself during this operation? In other words, if we draw the horseshoe right at the beginning, inside the top left-hand square of Figure 2.12, how will it be deformed? If we follow carefully what is going on step by step, we find that the initial horseshoe

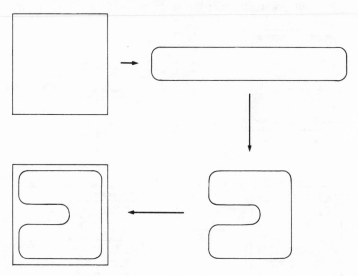

FIG. 2.12. Smale's horseshoe

is first elongated and then folded over, and that it ends with two branches in each of the two branches of the final horseshoe (Fig. 2.13). In other words, we get a kind of double horseshoe, with four branches instead of two, lying inside the initial horseshoe.

We may, of course, proceed. What happens to the double horseshoe? It gives a quadruple horseshoe (eight branches) lying inside the double horseshoe, which lies inside the initial horseshoe, which lies inside the square. And so on, indefinitely.

The strange attractor is the "infinite horseshoe" which you get in the limit. Note that all these finite horseshoes are included in each other: the strange attractor is precisely the set of points which are included in all of them. It has infinitely many branches, and looks like a curve which comes back infinitely often, trying desperately to fill a surface but not quite succeeding. It is neither a curve nor a surface, so it is hard to visualize, but it is certainly there. To find it, all one has to do is to plot the successive positions of any point in the square. One will eventually plot out a strange object in the square,

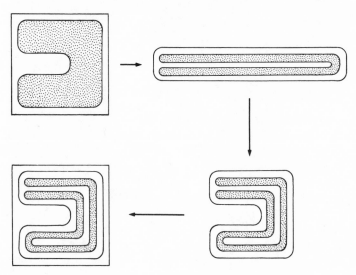

FIG. 2.13. The first iterate of the horseshoe

with a structure reminiscent of what we saw in the preceding section: it looks like a succession of layered strips, and under magnification each layer dissolves into another succession of layered strips. All these layers constitute the strange attractor for Smale's horseshoe, and the strange attractor for the Lorenz equations looks very much the same (with one more dimension).

In Appendix 2 we describe another approach to strange attractors, the Feigenbaum bifurcation. The interested reader will see how, by changing the equations of motion slowly and continuously, one can change a perfectly regular system with finitely many rest points into an horribly chaotic one, jumping around on a strange attractor. This phenomenon may be responsible for the onset of turbulence in hydrodynamics. We have all seen how smooth water flows break down as the velocity increases, and split into eddies of varying shapes and speeds. The equations, of course, depend on the ingoing velocity of the flow, and it may well be that, if the velocity becomes high, a strange attractor appears in the equations, thereby disrupting the behavior of the flow. This idea was put forward by Ruelle and Takens, and many scientists are working hard to vindicate it. This would be a major success of mathematical physics, since it would provide a mathematical model for turbulence, a physical phenomenon which has defied analysis for many centuries.

The qualitative approach is not a mere stand-in for quantitative methods. It may lead to great theoretical advances, as in fluid dynamics. It also has a significant advantage over quantitative methods, namely, stability. Anosov showed in 1961 that in systems of the Lorenz type, which are unstable with respect to initial conditions, perturbing the equations will amount to exchanging trajectories. In other words, each trajectory of the perturbed system will be close to some trajectory of the original system. These two trajectories may not have the same initial state: at time zero, as for any subsequent time, the positions will lie close to each other, but may be distinct. If we started the perturbed system at the same initial position as the original system, we would get a third trajectory, which would quickly run away from the other two, since the perturbed system is unstable with respect to initial conditions.

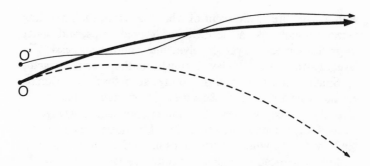

FIG. 2.14. Instability. Consider a trajectory starting from the
initial point *O* (*heavy line*). In unstable systems, a small
perturbation of the equations may be enough to change the
trajectory completely (*dashed line*). On the other hand, the
perturbed system will always have a trajectory which will stay close
to the first (unperturbed) trajectory, but it will have to start from
another point *O'* instead of *O*.

This shows that, even though every individual trajectory is
unstable, the overall pattern of trajectories may be stable. We
have seen that the weather is unpredictable. By this we mean
that, once general trends have been taken into account (succession of seasons, recurrence of the monsoon, presence of an
anticyclone near the Azores), we are left with fluctuations which
cannot be predicted beyond a few days. On the other hand, if
God were to build another earth, similar to but not identical
with ours (the shape of continents would be different, as would
sea depths and mountain ranges), the weather would not be
strikingly different from ours. We would recognize many of
the patterns we are familiar with: the same spring would come
over unknown blossoms, and the monsoon would bring rain
over strange lands.

In this kind of system, stability would be attached to the
overall pattern, and not to individual trajectories. Strange attractors, for instance, and the probabilities they carry, will not
be much affected by small perturbations. Think of die throwing. If we tamper with the dice, we will not change the possible
outcomes, which remain {1, 2, 3, 4, 5, 6}. The probabilities

will be altered, from $\{\frac{1}{6}, \frac{1}{6}, \frac{1}{6}, \frac{1}{6}, \frac{1}{6}, \frac{1}{6}\}$ to something slightly different. On the other hand, this tampering could dramatically affect the outcome of a particular throw, changing the result from 6 to 2, for instance. These dramatic changes, affecting individual trajectories, disappear into the overall probabilistic pattern.

This is yet another reason why we should use qualitative methods when studying this kind of system: they are the only methods which enable us to approach physical reality. Quantitative methods, even when feasible (this is, if the computations involved can actually be performed), are unrealistic, because their results can be applied to the system only when it is isolated from all outside influence. As we have seen, if the system is unstable with respect to initial conditions, the gravitational attraction of an electron at the boundary of the known universe quickly becomes a significant factor. The qualitative approach then is the only one which enables us to reach objects which are stable, that is, insensitive to small perturbations. The price to be paid is high: one must give up trying to predict the future in individual cases. The only predictions that can be made are short run. For the long run, one must resort to statistical methods.

In addition, the qualitative approach gives us insight into the system. Having identified a strange attractor, for instance, enables us to understand the patterns which the system will follow in the long run, even if definite predictions cannot be made. We now turn to another domain where the qualitative approach will show still more possibilities: catastrophe theory.

3
THE COMEBACK OF
GEOMETRY

A WORD OF WARNING

Catastrophe theory was born in 1972, with the publication of a very learned book by René Thom entitled *Structural Stability and Morphogenesis*. It has also a subtitle: *An Essay on the General Theory of Models*. Title and subtitle together indicate that here we have a mathematical theory with unusual claims to universality.

Thom is a leading mathematician; his ideas were already known in mathematical circles and spread very quickly to scientists in general and to the public at large. Catastrophe theory became an instant success, treatises and research papers were written, articles and interviews appeared in newspapers and magazines. In fact, so much has already been written about catastrophe theory that it seems quite superfluous to add yet another chapter to this growing list.

Despite so many explanations and commentaries, I still have the feeling that the instant success of catastrophe theory is largely due to a misunderstanding of what a theory is and what catastrophe means.

Let me first say what catastrophe theory is not. It does not announce catastrophes; it cannot tell whether the world will end in a nuclear war. Catastrophe theory does not enable us to make precise, quantitative predictions, the way relativity theory does, for instance. Neither can it be proved or disproved by an experiment, and so the question arises whether it is a scientific theory at all.

Actually it is—but it is much closer to biological theories like that of evolution than to physical theories like that of relativity. It fits certain facts together and provides an abstract setting to grasp them all at once. It is a way to make some sense out of the hopeless tangle of natural phenomena, like a listening device which picks some garbled message out of the overpowering background noise. It is a mathematical code to help us decipher the book of nature.

Let us hear what Darwin has to say on this subject:

> After my return to England it appeared to me that by following the example of Lyell in Geology, and by collecting all facts which bore in any way on the variation of animals and plants under domestication and nature, some light might perhaps be thrown on the whole subject. My first note-book was opened in July 1837. I worked on true Baconian principles, and without any theory collected facts on a wholesale scale, more especially with respect to domesticated productions, by printed enquiries, by conversation with skilful breeders and gardeners, and by extensive reading. When I see the list of books of all kinds which I read and abstracted, including whole series of Journals and Transactions, I am surprised at my industry. I soon perceived that selection was the keystone of man's success in making useful races of animal and plants. But how selection could be applied to organisms living in a state of nature remained for some time a mystery to me.
>
> In October 1838, that is, fifteen months after I had begun my systematic enquiry, I happened to read for amusement "Malthus on Population," and being well prepared to appreciate the struggle for existence which everywhere goes on from long-continued observation of the habits of animals and plants, it at once struck me that under these circumstances favourable variations would tend to be preserved, and unfavourable ones to be destroyed. The result of this would be the formation of new species. Here then I had at last got a theory by which to work.[1]

1. "Autobiography," in *The Life and Letters of Sir Charles Darwin*, ed. Francis Darwin (New York: Basic Books, 1959), 1:67–68.

No one doubts that evolution theory is a scientific theory. Even those who fight it in favor of a literal interpretation of Genesis concede this point. And yet it is widely different from physical theories, like that of gravitation. Newton gathers various facts, the motion of planets, the tides of seas and oceans, the fall of apples from trees, and explains them by a single mathematical law, which determines them completely. He does not really try to understand how gravitation works, and where this force of attraction comes from: how can a massive body make its presence felt instantaneously throughout the universe, as Newtonian physics calls for? What is the material support for this kind of action? These questions went without answer for three centuries, but this lack of physical understanding did not affect the mathematical model. During all this time, and even today, it has been used with great consistency and accuracy to study the relevant phenomena, predict their future, and reconstruct their past from present data.

Darwin's discoveries are of a different kind. He finds some hidden logic where his predecessors could see nothing but the Creator's will. He fits seemingly haphazard phenomena into one harmonious pattern, which connects them all together. But his theory is not predictive: he cannot tell which way evolution is going to take us. Darwin's famous law, "survival of the fittest," may determine the future of animal species, but it does not enable us to predict it, the way gravitation theory enables us to predict the future positions of planets.

The main contribution of evolution theory is to single out a central fact, the evolution of species, to which a host of biological phenomena are subordinate. It is also to provide us with ideas which will enable us to understand certain changes and transitions. Lamarck's main idea, for instance, is that organs develop in proportion to their use, and that characters which an individual has acquired in its lifetime may be transmitted to its descendants. Darwin, on the other hand, believes that the true motor of evolution is the struggle for life, with the resulting survival of the fittest. For both Lamarck and Darwin, species will ultimately be adapted to their environment, even though they differ on the means by which this will be achieved.

No one blames evolution theory for not being able to predict in which direction it will take us. What will our descendants be like one million years from now? No one asks, and curiously no one cares. The overwhelming interest seems to lie in investigating the past: what were our ancestors like? Strangely enough, there is no satisfactory answer to that question either, even with the help of field work. Human paleontology is still looking for the missing link, which would connect *Homo sapiens* to the genealogical tree of all animal species.

Catastrophe theory is a scientific theory, in the same way that evolution theory is scientific. It should not be understood as a predictive or quantitative theory, as physical theories are. Frequent misunderstanding is caused by the fact that catastrophe theory relies on deep mathematical results, pertaining to the classification of singularities. The Newtonian model immediately comes to mind, and one jumps to the conclusion that mathematics will be used in the same way, that is, to predict. However, as we have seen in the preceding chapter, the Newtonian equations of motion quickly become too complicated to handle, and quantitative predictions become impossible, so that one is forced to resort to qualitative methods. This is where catastrophe theory starts off: it is supposed to deal with cases when the defining equations are either unknown or unwieldy. It can only be qualitative. The best it can do is tell us what patterns of behavior to look for. The mathematics of catastrophe theory are no substitute for a detailed mathematical model of the system under consideration. Whenever such a model is available, or can be worked out from the physics of the system, it should always be preferred.

DISSIPATIVE SYSTEMS

We now turn to a very special category of dynamical systems, the so-called *dissipative* systems.

Dissipative systems have very simple dynamics: every motion tapers off toward some final rest position. The possible rest positions are called equilibria.

Let us go into this. A dissipative system may have one or several equilibria. If initially the system is set at an equilibrium position, with zero speed, it will never move away: this particular motion of the system is nothing but an infinite stay at this equilibrium position. For any other initial condition, if the system starts from a nonequilibrium position, or if it is set at equilibrium with nonzero speed, the system starts moving. But the motion eventually slows down, the oscillations taper off, and the system gets closer and closer to some equilibrium position, which it reaches in infinite time.

If we know where the equilibria of a dissipative system are located, we understand its dynamics: whatever the initial conditions (position and velocity) are, the system will eventually find its way to some equilibrium, and stay there. There can be no periodic trajectory, for instance, with the system going indefinitely through the same positions without ever coming to a rest. So planetary systems cannot be dissipative, as we observe from the presence of Keplerian orbits. Dissipative systems cannot have complicated trajectories, like the ones we saw in the preceding chapter: the only possible behavior is heading for an equilibrium position and staying there.

The damped pendulum provides a good example of such a system. Imagine a rod, hanging down from one of its extremities, with a metal sphere attached to the other extremity. We choose a rod instead of a string, so that large oscillations will be easier to observe.

We immediately find an equilibrium: the rod is vertical, with the sphere down. Indeed, if the pendulum is left in this position, with zero speed, it will not move away. There is also another, less noticeable, equilibrium, with the pendulum upside down: the rod is vertical again, but this time the sphere is up, on top of the swing. If the pendulum is left in this position, with zero initial speed, it will just stay there. On the other hand, the slightest deviation from the vertical position, or the smallest impulse on the metal sphere, will result in the pendulum falling away from this equilibrium, swinging back and forth in ever decreasing oscillations, and finally coming to rest at the other equilibrium. This makes the upside-down equilibrium hard to realize and to sustain experimentally; we call it an unstable equilibrium.

Now, if we start the pendulum at random, either by releasing the sphere from a nonvertical position or by giving it an initial impulse, we will observe the motion to be progressively damped, so that the pendulum eventually comes to rest at its stable equilibrium, with the sphere down: first a few complete turns, if the initial impulse was strong enough, then some large swings back and forth, with decreasing amplitude, evolving into smaller and smaller oscillations around the vertical position, which finally are damped into rest. This shows also that the vertical position, with the sphere down, is a stable equilibrium, and thus very easy to sustain experimentally: if by accident the system deviates from equilibrium, its natural motion will bring it back into position.

The damping of the pendulum is due to various kinds of friction, most notably air resistance. If this friction is increased—for instance, by operating under water, so that air resistance is replaced by water resistance—the oscillations disappear, and the pendulum will be seen to fall directly into its equilibrium position. Friction operates by changing the kinetic energy of the system into heat. Now kinetic energy is that part of total energy which the system actually uses to move. Friction converts it to heat, that is, transfers it to the outside, so that it is irretrievably lost for the system. When none is left, the system comes to rest: an equilibrium has been reached. This phenomenon is called "dissipation of energy"; hence the name we coined for dissipative systems.

The damped pendulum also tells us that we must distinguish between stable and unstable equilibria. Both are positions where the system can stay indefinitely, but only the stable ones can act as rest points for other trajectories, thereby focusing the dynamics of the system. This point will be made even clearer by a two-dimensional example.

Put a marble in a bowl. It will roll, and perhaps slide, along the edges. Eventually, the friction will cause it to rest at the bottom of the bowl.

Now let us make the bowl more complicated. We do not want it round anymore, we want it asymmetric: its bottom will have two basins, one slightly higher than the other, separated by a threshold (Fig. 3.1). If we now release the marble inside this new bowl, its motion will certainly be more complicated, but still it will come to rest at the bottom. There are two stable equilibria, which can be observed experimentally: they are the lowest points in each basin. There will also be an unstable equilibrium, somewhere on the threshold between the two basins. Ideally, if the marble is put at the right point on the threshold, it will just stay there forever, as if it could not make up its mind into which basin to fall. One should think of a mountain pass separating two valleys; at the top of the pass, a round pebble or boulder is in uneasy equilibrium, waiting for some outside force to push it one way or the other. This unstable equilibrium, at the threshold position, will not be observed experimentally, since the slightest perturbation will cause the marble to fall toward one of the two stable equilibria.

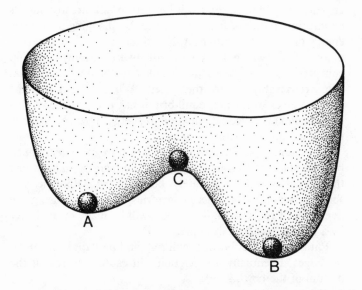

FIG. 3.1. The asymmetric bowl. The three equilibrium positions have been indicated. Two of them are stable (*A* and *B*), and one is unstable (*C*).

We can complicate the bowl further, and add more basins and more thresholds. The final shape one arrives at is some kind of mountain landscape, with basins separated by ranges and communicating by passes. One can plot out the crest lines on a map and indicate the bottom of each basin. On each crest line, summits and passes alternate: between two summits there must always be a mountain pass.

We can then understand the behavior of our marble in this relief by a natural analogy: the streaming down of rainwater. It streams down the slopes, to the bottom of the basins, where lakes form: these signal the stable equilibria. Two rain droplets, falling 1 foot apart but on different sides of the crest line, will trickle down to different lakes, many miles apart. These natural boundaries, the crest lines, are punctuated with peaks and passes, which are unstable equilibria.

This is a general picture of dissipative systems. One has first to define precisely what is meant by the state of the system. For instance, the damped pendulum and the marble rolling down its bowl are second-order systems, which means that the state is defined as position and velocity, not merely position. Once this is understood, the motion of a dissipative system is faithfully described by the streaming-down analogy. Each stable equilibrium will have its own basin of attraction, and every motion started in this basin will end up in the corresponding equilibrium position—just as rainwater fallen on the slopes will flow down to the lake. The various basins fill up the state space, except for the boundaries between them, which act like crest lines. These boundaries are punctuated with unstable equilibria, which have no great practical significance. Unless the initial point is located precisely on a boundary—an exceptional situation, which is unstable anyway—it will belong to the basin of some equilibrium, which will be the final rest point of this particular motion.

It is tempting—and it may even be realistic, if the motion is fast enough—to forget about the transitional states, and to see only the initial and final ones. The dynamics of the system then are given by the simple formula:

$$\text{initial state} \rightarrow \text{final equilibrium.}$$

This transition will not be continuous: small changes in the initial state may lead to different equilibria. Indeed, if the initial state is located near a boundary, a slight displacement will be enough to make it cross this boundary into another basin of attraction. Note also that initial states are transitional, while equilibria remain forever. Experimentally, except for brief transition periods when the system is knocked out of equilibrium, only equilibria will be observed.

So much for two-dimensional dissipative systems. For complex dissipative systems, the state space may have many more than two dimensions—ten, a hundred, a thousand, or even more. This means that ten, a hundred, a thousand, or more variables are needed if one is to pinpoint one state of the system, and this is realistic if one is modeling the kind of physicochemical systems that occur in a living organism. But our two-dimensional analogy, of water streaming down a relief, is still perfectly correct. The relief even bears a name: it is called the *potential* of the system. Each state of the system corresponds to a point on the map, and the height measured at that point is the value of the potential for this particular state.

Stable equilibria are the minima of the potential. This is only an elaborate way to express the simple idea that water will flow down the slopes and accumulate at the bottom. A minimum of the potential is the lowest point in some basin; a maximum is a peak on the ranges between two basins. We now switch to this new language. We will speak of potential instead of relief, and we will say that the potential decreases along the trajectories, instead of saying that the water streams down the slopes. This is really the same thing: if you look at the same droplet of water at two different moments, its altitude must have decreased, since it streams down. Mathematically speaking, if we now forget about the two-dimensional analogy, the state S_t at time t is completely determined by the state S_0 at time 0, through differential equations. If the value $V(S_t)$ of the potential at time t is to be compared with its value $V(S_T)$ at some later time T, it will be found to have decreased, that is, $V(S_T)$ is smaller than $V(S_t)$. If it is unchanged, that is, if $V(S_T)$

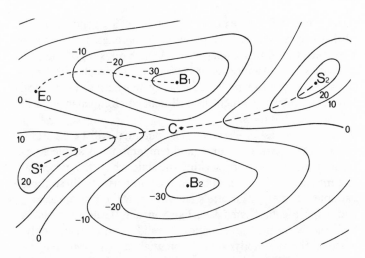

FIG. 3.2. The potential of a dissipative system, represented by its level lines. The system will flow down the slopes to one of its two stable equilibria, B_1 or B_2, at the bottom of the basins. If it starts precisely at one of the summits S_1 or S_2, or at the mountain pass C, it will remain in unstable equilibrium. Notice the crest line separating the two basins, and a typical trajectory starting from E_0.

is equal to $V(S_t)$, then S_T has to be an equilibrium, so that S_T coincides with S_t and the system has not moved.

The fact that the potential decreases along the trajectories has several interesting consequences, the most important being irreversibility. As soon as the system has left the initial state S_0 and reached a different state S_t, the value $V(S_t)$ will be smaller than $V(S_0)$, and there is no way that, in the subsequent evolution, it will ever regain its initial value $V(S_0)$. So the system will never come back to S_0. In other words, the system cannot go twice through the same state. This means, for instance, that there will be no periodic trajectories, which would force the system to go indefinitely through the same states.

Examples of dissipative systems abound, and the associated potential usually has a well-recognized physical meaning. Mechanical systems which lose their energy by friction, and which have no external source of energy to rely on, are good examples of dissipative systems, the potential being energy. Other systems turn out to be dissipative in a less obvious way. Thermodynamics, for instance, associates with physicochemical systems a variety of functions (free energy, free enthalpy, chemical potential, entropy) which, in well-defined circumstances, will be used as potentials.

Take a gas, as an example. The potential will be a certain function of pressure, volume, and temperature. A stable equilibrium will be found by seeking the state where this potential is a minimum. We then obtain a relation between these three variables, namely, Mariotte's law $PV = RT$ for an ideal gas, and a more complicated law, like van der Waals's law, for a real gas. Note that this relation is valid at equilibrium only, whereas the thermodynamical potential can (in principle) be evaluated at all possible states of the gas, including nonequilibrium, non-Mariotte ones.

But such states are never observed unless some special experiment is set up to push the system away from equilibrium. One could, for instance, open up to the gas an additional volume, or perturb the pressure locally by sound waves. One would then observe the nonequilibrium states during the brief transition period the system needs to regain equilibrium. During this period, the gas would not conform to Mariotte's law or any other equilibrium relation between pressure, volume, and temperature. Worse still, these three variables are well defined at equilibrium only; for transitional states, away from equilibrium, they do not really make sense. Describing the thermodynamic potential in terms of these variables, as I just did, is a gross oversimplification. To describe nonequilibrium states, a more detailed model is needed, using more variables to define the state.

The most detailed model available is Boltzmann's: the gas is but a collection of molecules, bouncing upon each other and upon the container walls. If we are dealing with about 10^{23} molecules (a realistic quantity of gas), this means that we have replaced a model with three variables by a model with 6×10^{23} variables (each molecule being characterized by its position and its velocity—six variables per molecule in all). In addition, Boltzmann's model does not conform to standard thermodynamics; in other words, it contradicts the simpler, three-variable model. Indeed, Boltzmann's model is not dissipative anymore; there is no potential willing to decrease along the trajectories, and there are even periodic solutions! We saw one of these periodic solutions in the preceding chapter, when we studied Poincaré's recurrence theorem. We saw how a gas, which had expanded to fill the available volume, was supposed to contract itself back into its initial place, over and over and over again. Clearly such a detailed model is not realistic either. The model to be chosen for each situation must be somewhere between the thermodynamical model, to be used for equilibrium and near-equilibrium states, and the Boltzmann model.

As far as thermodynamics are concerned, we must keep in mind that the dynamics which lead the system to equilibrium and keep it there are not described adequately by the model. The three thermodynamical variables are well defined at equilibrium, and not otherwise: pressure and temperature may be defined locally, but not for the whole gas, unless it is at equilibrium. The thermodynamical model gives an adequate description of the states of equilibrium, but fails to provide a relevant picture of the way nonequilibria converge to equilibria. It is a theory of statics, not of dynamics.

Here again we see that in dissipative systems, only the equilibria are important, and all the transitional states, with their underlying dynamics, may as well be forgotten. This simple observation is derived immediately from experimental facts: real dissipative systems are mostly observed at equilibrium, hence at rest. But it has far-reaching consequences. Static phenomena may now be provided with dynamical models: we will argue that the system is dissipative, and that the dynamics may be forgotten in favor of the equilibria. Little does it matter now if the potential which the model ascribes to the system seems to be more a mathematician's fiction than a part of physical reality. After all, entropy or enthalpy is not accessible to direct measurement either, at least not as readily as temperature. And if the dynamics of the model are wrong, that is, if it gives an unrealistic picture of the transitions to equilibrium, that also does not matter too much.

For instance, when the thermodynamical model fails, there are more refined models to resort to, such as the Boltzmann model. If we really wish to study gases away from equilibrium, we can always build an intermediate model which has the required accuracy, so that the thermodynamical model is seen as a limiting case.

This possibility is not always open. It is very tempting to decide that some phenomenon we do not understand is modeled by a dissipative system, without any kind of empirical or theoretical evidence. One simply comes up with a potential that has as many minima as the phenomenon to be studied has stable forms, and postulates some vague relationship between the former and the latter. This approach verges on wishful thinking, and there is not much substance to it, unless there is more evidence of the fact that there should be a dissipative model and of what its potential should be like. Unfortunately, some of the most popular applications of catastrophe theory are in behavioral or social sciences, where it has been used to model everything from mutinies in prisons to attack patterns of dogs. There is no theoretical justification to such models; they are at best an educated guess from a mathematician.

CATASTROPHES

There are, however, quite a few natural systems which can be modeled by a dissipative system with some degree of realism. The mathematical model—insofar as it can be written down completely—is a potential, depending on a large number of variables. These variables, which we henceforth call *internal*, define the state of the system. The potential regulates the evolution of the system by means of the internal variables. If the system is complex, the complete description of one state will require an enormous number of internal variables, all of which will be used in writing down the potential. We need not do this; never mind the complete mathematical expression of the potential—all we need to know is that it exists. Its minima are stable equilibria, and the system will usually be found resting at one of them.

We now act on the system from outside. To be precise, say that, in addition to the internal variables, the system also depends on a few more variables, which the observer can modify at will; these new variables are called *external*. We will also refer to external variables as parameters. The potential will depend both on the internal and on the external variables. The observer cannot reach into the system and tamper with the internal variables, but he can adjust the external variables. Doing so amounts to modifying the potential of the system, and thus changing its equilibria.

To understand this, it may be best to return to our former analogy comparing the potential to a landscape. To change the external variable is to change the relief. The crests may be heightened or eroded, the valleys may be deepened or filled up. The shape of the basins will be altered, and their lowest points will move. If there is a lake at the bottom, its position on the map will change. The boundaries between basins will also be modified, and a creek that went into one lake may henceforth end up in another.

Slow, progressive modifications of the relief may result in swift, abrupt changes. If, for instance, two basins, one higher than the other, are separated by a mountain pass, and if this threshold is lowered continuously, there will be a point when the first basin is engulfed by the second. Imagine a lake at the bottom of the first basin (Fig. 3.3). If we lower the altitude of the pass connecting the first basin to the second, the lake will

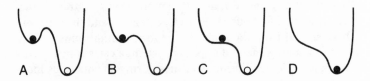

FIG. 3.3. The position of the marble during the *ABCD* transition is indicated in black. The position during the backward transition *DCBA* is shown in white.

not be affected, until the pass is lowered to the altitude of the
lake itself. Then the boundary between the two basins vanishes,
the upper basin merges with the lower basin, and the lake
disappears.

There is a critical value for the height of the pass: it is the
altitude of the lake itself. As long as it is not reached, modifying
the height of the pass has but little effect. If, however, it is
crossed in either direction, a substantial change takes place,
that is, a lake appears or disappears.

This is what Thom calls a *catastrophe:* a major change in
the system, consecutive to a minor change in the external
variables.

Let us simplify this example even further, and make it one-
dimensional. We are now dealing with a system with a single
internal variable; let its potential be the one of Figure 3.3,
position *A*. Note that there are two minima for the potential,
that is, two stable equilibria for the system. Change this potential slowly, by lowering the height of the threshold separating the upper and lower basin. The main stages are recorded
in Figure 3.3, positions *B–D*. Nothing much happens until
stage *C* is reached: this is when the critical height is attained.
The two basins merge, the upper equilibrium disappears, and
we are left with a single equilibrium and a single basin.

All this can be made more visible if a marble is put in the
upper basin at the start. As the deformation proceeds (that is,
as the external variables change), the marble will stay put until
position *C* is reached. It will then fall into the lower basin. If
we now go to the same stages in reverse, the marble will not
climb back to its initial position—the lower basin cannot empty
into the upper one. So, after a complete cycle of transformations, we will find ourselves back at stage *A*, and the state will
have shifted from the upper to the lower equilibrium.

For a more general dissipative system, a catastrophe occurs
when a stable equilibrium vanishes during a continuous modification of the external variables.

The precise mathematical expression of the potential is not
required; we do not even have to know the number of internal
variables. All we need to know is that the system is dissipative,
so that there is indeed a potential regulating the internal vari-

ables. As far as we, the observers, are concerned, the system is a black box, closed to inspection. We will not describe its inner workings. We will just try to record its response to external stimuli. This will amount to a description from the outside, a purely phenomenological one. After all, the system is nothing but the totality of all its possible responses to the world outside. In other words, we attempt a description from the outside, instead of an understanding from the inside.

Of course, the external stimuli must be carefully monitored. It is essential for catastrophe theory that a small number of external variables be chosen, say three, which will be the only ones to vary, all other parameters being kept fixed. The values given to these three external variables will then be registered as a point in three-dimensional space. If these values are changed, the corresponding point moves along, and the potential of the system gets modified in some (unknown) fashion. The system, initially at a stable equilibrium, follows the latter in its variations. It may happen that, for certain values of the parameter, this particular equilibrium vanishes. The system then jumps to another stable equilibrium; that is, we get a discontinuous response to a continuous change in the external variables. The set of parameter values where this happens is called the catastrophe set. It is a kind of boundary which stretches across three-dimensional space: if the point representing the values of the external variables crosses this boundary, the system will jump from one equilibrium to another. This change of equilibrium will be perceived by the observer as a sudden and substantial change in the properties of the system—a phase transition, for instance, like water solidifying into ice. If one lowers the temperature of water regularly, nothing much happens, until the critical temperature of $0°$ is reached, at which time ice appears. Here there is just one external variable, temperature, and one catastrophic value, $0°$ C. The parameter space is one-dimensional, and the catastrophe set consists of one point, 0, which acts as a boundary between negative temperatures (ice only) and positive ones (water only).

Zeeman has constructed a two-dimensional catastrophe machine. It is a flat wheel, rotating freely around its axis. Two rubber bands are attached to the wheel at some point M: each of them has one end pinned down at M and one end free. We take one rubber band, stretch it, and fix its other end somewhere in the plane supporting the wheel, say at P. The wheel will then find an equilibrium position, with the distance MP at minimum. We now take the free end of the second rubber band and pull it: the wheel will find a new equilibrium position, depending on the position H of the hand holding the rubber band.

There is no doubt that the system is dissipative: there is a single internal variable, namely, the angle by which the wheel has turned, and it is quite elementary to compute its potential. This potential depends on the position of the hand on the plane supporting the wheel. In this case the parameter space is two-dimensional, and the point H represents two external variables. We will now move H across the plane, that is, move the hand holding the free end of the second rubber band.

This catastrophe machine is easy to build, in plywood or in cardboard, and quickly enough one discovers a strange region in the plane, a kind of curved lozenge with pinched summits: its boundary is the catastrophe set. If it is crossed from the inside of the lozenge to the outside, a catastrophe happens: the wheel suddenly spins one half-turn and finds itself a new equilibrium position! If we now reverse tracks, and cross the boundary again at the same point, going from the outside to the inside, nothing happens: the wheel just smoothly follows the hand's motion. So the catastrophe set may be crossed freely in one direction; if it is crossed in the other direction, however, a catastrophe will happen.

In fact, whenever H is outside the lozenge, there is only one possible position for the wheel, but if it is inside there are two. So, if the hand holds the rubber band inside the lozenge, which of the two possible positions the wheel assumes will depend on what particular route the system has traveled to reach that position. If, for instance, H travels into the lozenge and out, say from Q_1 to Q_2, and then backtracks from Q_2 to Q_1, the hand will find itself going through the same points, with the wheel in different positions. The system takes into account

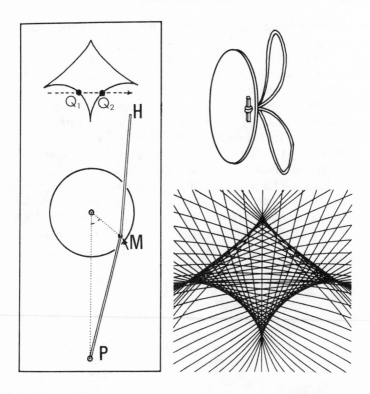

FIG. 3.4. Zeeman's catastrophe machine is shown on the left-hand side of the picture. The wheel swings around its center, which is fixed on a plane. A rubber band is pinned down at P, attached to the wheel at M, while the loose end H is held in the experimenter's hand. Some experimenting will readily discover a region in the plane where the wheel swings back a half-circle. For instance, if the hand holding the rubber band travels along the dashed line, the wheel will swing back when the point Q_2 is reached. Going back along the same path, but in the opposite direction, one will observe a jump at the point Q_1. There are four cusps to be observed in the plane. On the right-hand side, it is suggested how the rubber band may be attached to the wheel, and a computergraphic shows the catastrophe boundary in more detail.

previous history: it remembers, albeit in a very rudimentary fashion.

The Zeeman machine is a dissipative system with a single internal variable, and two external variables. Catastrophe theory states that similar features will be observed for almost any kind of dissipative system, with hundreds of thousands of internal variables, provided that one acts on two external variables only. In other words, one decides on two parameters on which to act. Catastrophe theory then provides a general model for this situation: it is the *cusp*, depicted in Figure 3.5.

This is a three-dimensional picture. Downstairs we have a horizontal plane, above which lies a folded surface, which is shown cut by two vertical planes, one forward and one backward. The fold ends at a point A, which projects down to a

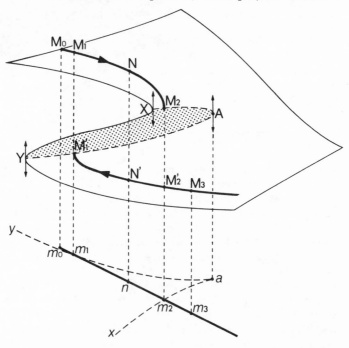

F I G . 3.5. How to cross the cusp. The shaded area represents unstable equilibria.

on the horizontal plane. The contour XAY of this fold (imagine the sun being vertical and projecting shadows downward) projects down as a cusp xay, with summit a. For every point inside this cusp, there are three points of the surface lying directly above it, while there is only one for any point lying outside the cusp.

The horizontal plane will be our space of parameters: it is two-dimensional. The third dimension in our drawing corresponds to the internal variables: one is enough, although the system may have many more. The folded surface then is the set of equilibria. The theory shows that the shaded part bounded by XAY consists of unstable equilibria. They have no practical meaning, and may as well be cut out. Once this is done, every point in the horizontal plane has one (outside the cusp) or two (inside the cusp) points of the (cut) surface projecting down to it. In other words, for every pair of parameter values, there are one or two stable equilibria. The height of the surface above the plane gives the value of the internal variable.

Let us now change the values of the external variables, that is, move a point m on the horizontal plane. Start from an initial position m_0 outside the cusp: corresponding to m_0 there is a well-defined equilibrium M_0. Now move towards the cusp; we cross a first boundary, ay, at the point m_1, and the corresponding equilibrium M_1 moves smoothly along the upper sheet. Continuing in the same direction, we meet the second boundary, ax, at the point m_2, the corresponding equilibrium being M_2, where the upper sheet drops off. If the parameter values now move beyond m_2 to m_3, say, the state cannot follow on the upper sheet: it must jump down to the new equilibrium M_3 and continue on the lower sheet.

If we backtrack along the previous path, there will be no catastrophe this time when crossing ax: the system just proceeds smoothly along the lower sheet. At the point n, for instance, the observed equilibrium will be N', whereas it was N when we went through the first time. When we reach m_1 again, and cross the boundary ay in the other direction, a catastrophe occurs: the state jumps over to the upper sheet, and we end up at M_0, with parameter values at m_0, as we began.

In this case, the catastrophe set consists of the two lines ax and ay, joining in a cusp. If they are crossed from the inside to the outside, a catastrophe will occur. Note that if we go from m_1 to m_0 by avoiding the catastrophe set altogether (just make a detour around a), the transition will be perfectly smooth as the upper and lower sheet merge gently beyond A.

THEORY

Catastrophe theory teaches us that, if one acts on a dissipative system via two external variables, all other parameters being kept fixed, the catastrophe set will be a collection of cusps in the plane. This statement tells us nothing about the inner workings of the system under consideration. It does not even tell us whether catastrophes will occur at all; quite possibly, the system will respond smoothly to all variations of the external variables, and the catastrophe set will turn out to be empty. It tells us that if catastrophes occur, the boundaries along which they happen will be smooth lines, punctuated with cusps. A typical example is the catastrophe set in the Zeeman machine, which we now recognize as a collection of four cusps.

Catastrophe theory does not say anything further about the catastrophe set. There may well be no cusps at all, the catastrophe set then consisting of smooth curves without corners. Catastrophe theory simply states that no shape more complicated than a cusp will ever occur in the two-dimensional case. One could have thought, for instance, that catastrophes would occur at isolated points in the plane, or that there would be a whole region, like the interior of a triangle, consisting only of catastrophe points.

Catastrophe theory tells us that this does not happen *in general.*

Beware of the provision "in general." This is the Achilles heel of the whole theory. Catastrophe theory does not claim that its statements hold true for all dissipative systems, but only that they hold for most such systems. It may happen that a certain dissipative system does not conform to the conclusions of catastrophe theory, so that, for instance, all points in a two-dimensional area are catastrophic. The theory then says that, if we were able to reach into the system and modify its potential ever so slightly, we would obtain a new system, the catastrophe set of which consists only of smooth boundaries and cusps.

This is, of course, a strange idea. We are in the habit of studying equations as we get them from the model, not of modifying them to suit our whim. Nature will not change the potential of the system just to make our task easier, or to uphold catastrophe theory. On the other hand, if almost all potentials conform to the theory, why would Nature have to pick an exception? If such an exceptional potential does occur, there must be a hidden reason, an underlying symmetry, for instance, which should be interesting to discover. So the merits of this provision in catastrophe theory, that it holds only for most dissipative systems, are not a clear-cut issue. They have been a matter of debate for a long time now, and we shall have more to say about it.

Catastrophe theory gives similar statements when there are more external variables: three, four, five, or six. Six is the magic number, above which the theory fails. If one is trying to act on seven external variables at once, the catastrophe set becomes too complicated to be broken down into simpler pictures: there are no pervasive shapes, like the cusp, to look for anymore. On the other hand, for six external variables or less, there are seven basic shapes, known as the "elementary catastrophes."

Let us sketch, for instance, what happens in the case of three external variables. The parameter space then is three-dimensional, and the catastrophe set must consist of smooth surfaces, which may be folded into one or more of the following shapes: the swallowtail, the wave (hyperbolic umbilic), and the hair (elliptical umbilic).

These names were coined by Thom (the terms in parentheses are for mathematical consumption). A glance at the picture is enough to see why. We may recall that the swallowtail also belongs to the carpenter's vocabulary, as a way of cutting beams to fit them together. The hyperbolic umbilic is seen as a long

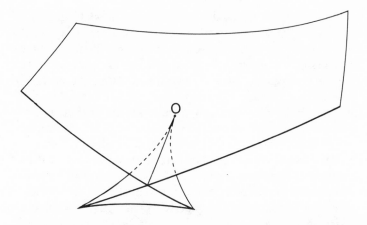

FIG. 3.6. The swallowtail

wave starting to break at its center; actually, the correct mathematical picture would require that we consider also the same wave breaking in the other direction.

These are the "elementary catastrophes" in dimension 3, as the cusp was the elementary catastrophe in dimension 2. In dimension 4, there are two more, the "mushroom" and the "butterfly," and the list is then complete. They serve as building blocks for the catastrophe sets. Their properties are those we already know from the cusp. Boundaries can be crossed freely in one direction (the system responds smoothly to the change in the parameters), but crossing them in the other direction induces a catastrophe (the prevailing stable equilibrium disappears, and the system jumps to another one). There are two possible equilibria in the region inside the swallowtail, none in the half-space lying above it, and a single one in the remaining region.

We can play games with these pictures to test our understanding of catastrophe theory. Assume, for instance, that we are acting via three external variables p_1, p_2, p_3 on a dissipative system which happens to have the hair (elliptical umbilic) as a catastrophe set. We then obtain in three-dimensional space a folded surface which looks like Figure 3.8.

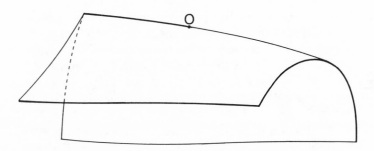

FIG. 3.7. The hyperbolic umbilic, or one-half of it (a full umbilic consists of this picture, plus its mirror image with respect to a vertical plane going through *O*). Notice that the vertical sections bend more and more from right to left, until they start folding at *O*.

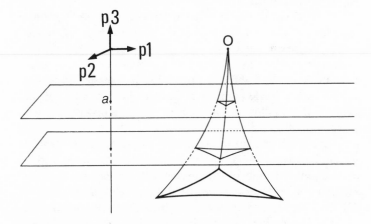

FIG. 3.8. The elliptical umbilic

Now leave the third external variable p_3 pegged at some a, and vary the two others, p_1 and p_2. We are now experimenting with two parameters only, p_1 and p_2, so that the relevant parameter space is the two-dimensional plane $p_3 = a$. The catastrophe set we will observe by restricting ourselves to p_1 and p_2 is the intersection of the umbilic with the horizontal plane $p_3 = a$. As we see from the picture, there are several possibilities, depending on the value a chosen for p_3. If we choose a, the height of the horizontal plane, so that it goes through the summit O of the umbilic, this intersection will be a single point, namely, O. But this seems to contradict catastrophe theory: in this two-dimensional (p_1, p_2) situation, we should be seeing smooth curves and cusps—not an isolated point all by itself.

Again, the answer is that catastrophe theory does not hold for all dissipative systems, but does hold for most of them. True, the conclusions of catastrophe theory do not hold if we peg p_3 at this very particular value a. But they become valid if we move a ever so slightly in any direction. Move it up, and the plane does not meet the umbilic any more; the catastrophe set disappears. Move it down, and the intersection is a curved triangle, the smooth boundaries of which meet in three cusps, according to the theory.

In this way, we see that the sharp edges of the hair (and the edges of the swallowtail and the wave as well) are cusps, which can be observed even with two parameters. The smooth surfaces bounding the hair are folds, where one stable equilibrium disappears: the corresponding catastrophe can be observed with a single parameter. Only at the summit O of the hair (or the swallowtail or the wave) do we observe a truly three-dimensional phenomenon.

I have one last point to make. The shape of the catastrophe set can be changed in a purely artificial manner, for example, by changing the units of measurement. If larger units are chosen, this will lead to lower measured values, and all the pictures will be scaled down. One could imagine more complicated devices, like plotting a two-dimensional picture on a rubber sheet and then pulling the sheet. A point will remain a point, but a straight line will be twisted into a smooth curve. Rectangles and circles will become unrecognizable, but the elementary catastrophes will retain the same shape. No rescaling or twisting, however complicated, can create or destroy a cusp: the two boundary lines may be deformed, they will always meet tangentially at the summit, and we will recognize the cusp.

To conclude, let us reassert that catastrophe theory does not tell us a priori what will happen in any given situation. Even if we are dealing with a nice dissipative system, and we have three external variables at our disposal, there is no way we can guess whether we will observe a swallowtail or an umbilic. We might also observe something quite different, which would mean that the potential we are dealing with is exceptional. Only by experimenting can one answer these questions—or by computing the potential, if this is possible. And if it is a swallowtail after all, catastrophe theory cannot tell us anything about its location or its size.

The main contribution of catastrophe theory lies in the idea of studying a complicated system from the outside, by logging in its responses to a well-defined set of stimuli. It lies also in the interest directed toward qualitative changes in the system, e.g., sudden transitions from one state to another or the influence of past history on present state. Finally, it lies in the seven elementary catastrophes (of which I have only described five)—seven beautiful figures in geometry, to be carried around in one's mind and to be matched against what we observe around us.

A CRITIQUE

Catastrophe theory was one of the major scientific events of the seventies. As soon as Thom's book appeared, and despite its very technical content, the interest in catastrophe theory spread far beyond the scientific community, and the mass media carried laudative papers and interviews. It is quite unusual for a mathematician to enjoy this kind of popularity. Thom's earlier work, considered by professional mathematicians to be much more interesting and difficult than catastrophe theory, won him the Fields medal in 1962, which is the highest award in the field, since there is no Nobel Prize for mathematics. But Thom's reputation remained strictly confined to mathematical circles until the birth of catastrophe theory.

Christopher Zeeman, an English mathematician, became the most active proponent of catastrophe theory by systematically seeking applications in a variety of down-to-earth situations: heartbeats, attack patterns of dogs, mutinies in English prisons. Thom himself applied catastrophe theory to morphogenesis (how an animal embryo develops from a single cell—the fertilized egg) and to the study of language. The relevance of these models was disputed, and the scholarly debate quickly developed into a polemic.

At times it felt like a comedy. Some overzealous catastrophists would boldly call a "cusp" any shapeless cluster of experimental points, and would be held up to ridicule and contempt by their opponents. In a series of papers, the Argentine mathematician Hector Sussmann attacked catastrophe theory on the grounds that it did not have any reliable application. As he puts it, "In mathematics, names are free. It is perfectly allowable to call a self-adjoint operator an elephant, and a spectral resolution a trunk. One can then prove a theorem, whereby all elephants have trunks. What is not allowable is to pretend that this result has anything to do with certain large gray animals."

To be sure, the word "catastrophe" conveys more than it means, and has led people to expect from catastrophe theory much more than it can actually deliver—which never was very much. In the twenty years of its existence, there has not been a single undisputed success of catastrophe theory in the field of experimental science, that is, an undisputed fact that could be explained more adequately by catastrophe theory than by other means.

There is, however, much more to say. In the first place, one should bear in mind that catastrophe theory is but a substitute for a better theory that was never born. What Thom really wanted was a theory that would account for all dynamical systems, dissipative or not. Such a theory would have described many more types of catastrophes than the disappearance of a stable equilibrium. For a nondissipative system need not go to a rest point: it can keep moving indefinitely, by running along a periodic orbit, or by approaching a strange attractor. A general catastrophe theory (as opposed to the elementary catastrophe theory of Thom) would face the formidable task of describing the possible transitions between these various stable motions.

We are very far from such a theory. The only thing that is relatively well understood is how a stable equilibrium turns into a stable periodic solution (Hopf bifurcation). We also know that certain transitions are so complicated that they cannot be described by a finite list, like the seven elementary catastrophes. The full repertoire is much larger, infinite in fact. So it is very doubtful that we will ever have a general catastrophe theory.

Thom's elementary catastrophe theory thus will remain, in all likelihood, the only tool at our disposal to describe the influence of external variables on complicated dynamical systems. It applies to dissipative systems only, and not all of them at that, so that one never knows whether it will hold for the particular system one is interested in. In view of all these restrictions, it is not surprising that well-authenticated examples are few and far between, except for machines built for that purpose.

But Thom's basic ideas turn toward metaphysics rather than science. In his book he explains that the shapes which we see in nature are basically catastrophe sets, so that the seven elementary catastrophes, and the general catastrophes if we only understood them, are the bricks which nature uses to build its infinite variety of shapes and forms. Two thousand years ago another mathematician, Plato, in the *Timaeus*, also tried to reduce the manifold shapes of nature to a few elementary ones. At this time there were only five regular solids known to mathematics, and Plato associated them with the basic elements of nature in the following order: the tetrahedron with fire, the cube with earth, the octahedron with air, the dodecahedron with the universe, and the icosahedron with water.

Plato's approach seems meaningless, and slightly ridiculous, to us nowadays. We fail to realize to what extent the Greeks' five regular solids have shaped our perception—the way we see the world around us. When a child piles up cubes, he is not only playing: he is learning what three-dimensional space is. In fact, he is generating the infinite, empty, Euclidean space:

it is empty so that there is room for cubes, and it is infinite so that there is no end to the number of cubes you can put in it. This Euclidean space, which our mind fills with cubes, is the framework of Newtonian physics, and it remained unchallenged until the twentieth century. In art one also finds this fascination with polyhedra. The cubist school in modern painting, for instance, has striven to disclose the hidden geometry of cubes which is buried under the shapes which surround us.

Thom is calling for a renewal, or at least an updating, of the unspoken geometry in our minds. Over the gates of Plato's utopian city, an inscription read "Let no one but geometers enter here." Plato and Thom explored the major problems of science as they knew it with the means that mathematics put at their disposal. In Plato's time it was cosmology; nowadays it may be biology. Plato saw the divine craftsman, the *demiourgos*, constructing the world by combining the five regular polyhedra in mysterious ways. Many centuries later, Kepler, seeking an explanation of the various sizes of the planetary orbits around the sun, would discover that if one tries to fit the five regular polyhedra into each other, their respective sizes, from the inner one to the outer, will be those of the planetary orbits, from Mercury to Saturn. Thom turns his attention to biology, and asks us to forget the five Platonic solids and Euclidean geometry. The seven elementary catastrophes are the first few words in the language of nature.

Each object in the world (as Thom sees it) is associated with a dynamical system. The shape, or form, of this object for an observer is nothing but the catastrophe set associated with this dynamical system. The object itself lies in the parameter space, usually three-dimensional Euclidean space, but it is not clear at all where the dynamical system lies, or even if it has a physical reality. For instance, the shape of a breaking crest wave does call to mind a hyperbolic umbilic, but no one has ever been able to build a dissipative potential for the wave from hydrodynamic considerations. Thom himself is not very concerned with this problem. He seems to be content, either with a Platonic attitude of having the dynamical system happen in some supernatural world or with locating it in the neurological phenomena which carry the perception to our brain.

These views are certainly more acceptable in the biological world than in the physicochemical one. We can easily accept that in any living tissue myriad simultaneous chemical processes take place each moment and determine the state at every point. The idea of a morphogenetic potential comes from the biologists themselves. Morphogenesis is the transformation of a single cell, the fertilized egg, into a differentiated embryo. With each point in the undifferentiated tissue we associate a potential, the internal variables of which probably somehow reflect the level of the various chemical processes going on at that point: this is the morphogenetic potential. It is a working assumption, neither supported nor contradicted by available evidence, which permits one to interpret the formation of the embryo in terms of catastrophe theory. The external variable is simply the position of a point inside the tissue. When two parts of the tissue differentiate, the boundary between them is seen as a fold catastrophe. Thom's book shows very interesting pictures of an egg at various stages. We see the tissue developing a groove, an exfoliation, or a needle, which will be understood in terms of cusps, swallowtails, or umbilics.

There is one last thing I ought to mention: the role allotted to time in the theory. For mathematicians time is but the fourth external variable, t, the three others being the coordinates which denote the position in space, x, y, and z, and we will describe catastrophes in four-dimensional parameter space. Physicists or biologists, on the other hand, will not observe this four-dimensional catastrophe directly. What they will see is a succession in time of three-dimensional catastrophes—that is, a catastrophe set evolving with time.

There are two elementary catastrophes in dimension 4: the fountain (parabolic umbilic) and the butterfly. Each of them corresponds to a succession of transformations of swallowtails and umbilics with precise rules—a beautiful dance which these geometrical figures perform, exploding out of a point, merging into each other, and leaving the stage.

If we cut a swallowtail by a plane going through its summit, all we observe is a smooth curve. If now we move this plane, the center of the picture explodes into a curved triangle, thereby revealing an unsuspected complexity: the swallowtail is seen as a succession of two-dimensional figures. In the same way, the fountain and the butterfly can be seen as a succession of three-dimensional figures. At the initial time, $t = 0$, one will observe a simple form, which will explode from its center into more complex shapes at subsequent times. This is the process by which, according to Thom, the embryo develops from the egg. It is a centrifugal process, consisting of successive differentiations from an organizing center, in contrast to a centripetal process, whereby more and more complicated organs would be added to a common body.

Catastrophe theory is a way to look at the world. In the fourth century B.C. Heraclitus wrote that war—*polemos*—was the father of all things, and that nature was the ever-changing battlefield between opposites. In today's language this may amount to precisely what Thom is saying: every form arises as the catastrophe set associated with some dynamical system. The disappearance of one stable equilibrium in favor of another—isn't this the outcome of a conflict where the stronger opponent kills the weaker one? This interpretation in terms of dynamical systems is supported further by the fact that Heraclitus saw the world as flowing and changing continuously: "Those who step down into the same rivers, bathe in the flow of an everchanging water,"[2] he said, or, as Plato reports, "No one can step twice into the same river,"[3] for the river flows, and the molecules which make up the particular water we stepped into are long gone when we enter the river the second time.

It is a miracle that a twentieth-century scientist is able to free himself from so many intellectual bonds and look at the world as if he were the first man to do so. His mood is wonderment, and he makes us share it with him: "After all, it probably is largely a matter of taste whether certain phenomena

2. Fragment 15.
3. *Cratylos* 402A.

will be considered to be scientifically interesting. Nowadays physicists build gigantic machines to study states, the lifespan of which does not exceed 10^{-23} seconds. It is probably right to use all the technical tools at our disposal to inventory all physical states accessible to experimentation. Another question, however, may legitimately be asked. A considerable number of everyday phenomena, so familiar that they go unnoticed, are quite difficult to account for: the cracks on an old wall, the shape of a cloud, the fall of a dead leaf, the foam in a glass of beer. . . . It would perhaps contribute more to the advancement of science if small phenomena of this kind were to be subject to a careful mathematical analysis."[4]

If we do not want to follow Thom on this particular ground, what is left of catastrophe theory? As we have seen, its immediate contribution to science is far below the hopes that were entertained in the pioneering days, the reason probably being that there are few well-authenticated dissipative systems, most natural dynamics being infinitely more complex.

On the other hand, catastrophe theory makes us aware of the transformations which are taking place in science. It may be a prototype of future theories, which will be qualitative rather than quantitative. It heralds the comeback of geometry, the new preeminence of pictures over computations.

The central fact has been analyzed at length in the preceding chapter: most deterministic systems are impossible to predict because their dynamics are too complicated for any meaningful computation to be possible. This leads us to the idea that some kind of qualitative knowledge should be possible, which would seek to inventory the possible outcomes rather than predict them.

This is precisely what catastrophe theory does, in its own restricted domain. By concentrating on the simplest dynamical systems, the dissipative ones, it provides a coherent mathematical model for certain deterministic systems which may be dubbed creative: they do not repeat themselves (memory); they create forms (morphogenesis).

4. *Stabilité structurelle*, p. 26.

The price to pay is time. Time is absent from catastrophe theory, which captures only a lifeless image. These beautiful figures with poetic names, the mushroom, the butterfly, are like marble statues in a crystal palace, outside which true time roams, free and alive. It has been cast out right at the beginning, as soon as we decided to forget the internal dynamics of dissipative systems and to retain only their stable equilibria. We have thereby rejected dynamics in favor of statics. Time will be seen only as the fourth dimension of space-time. In other words, it is reduced to a fourth dimension of space, and most of its fundamental properties, like irreversibility, are lost in the process.

This geometric image, a static reflection of an irreversible and fleeting time, brings to mind another one: Kepler's ellipse. Thom's elementary catastrophes, as well as Kepler's ellipses, attempt to reduce time to space and to understand it through geometry. Whereas Kepler uses the mathematical tools inherited from the Greeks, Thom has the benefit of modern differential topology. Kepler uses Apollonius's *Treatise on Conics*. Thom uses singularity theory. However, whereas Kepler's model leads us into Newton's world, which is closed upon itself, catastrophe theory is a glance into an open universe. In Newton's world there is no past and no future, since everything is determined by today's data. Time holds no surprise in store for whoever can handle the computations. In Thom's world the future is mostly hidden, and the mathematician inspects the flow of events for forms to recognize and classify, like a butterfly catcher.

4

BACK TO THE
BEGINNING

We have reached the end of our journey. Our starting point was Ptolemy's universe, a complex and sophisticated combination of circular motions. We left this construction, much weakened by age and by Ptolemy's overzealous successors, for the strength and simplicity of Kepler's universe. We saw Kepler's three laws regulating the planetary orbits. We then entered Newton's world, ruled by gravitation to the finest detail. This is a transparent world: the past and the future lie within our reach, buried in the information available today. It is the task of scientists to sift this information with the tools mathematics provides, and to reconstruct the state of the world yesterday or tomorrow from its present state. We can turn clocks backward or forward at our pleasure. Time can be reversed, like the three space variables, and the dynamics of Newton's universe are treated as problems in four-dimensional geometry.

Today, general relativity is the direct heir of Newtonian cosmology. The geometrical properties of Einstein's four-dimensional space-time translate into laws of motion. If we could see the four dimensions all at once, as the theory does, we would just see static objects in space-time. In reality, we see a succession of three-dimensional sections, which we understand as an object moving in space. Of course, from Newton to Einstein the mathematics has become infinitely more complicated. But the outlook is the same: time is absorbed into space; the laws of motion become problems in geometry. The universe is closed upon itself, regulated by strict determinism. The world is like a book, where all the past and the future are inscribed. Nothing unexpected can ever happen, if one knows how to read it.

The understanding of dynamical systems that we have today stresses the shortcomings of this outlook. As Poincaré was the first to point out, a detailed mathematical study of deterministic motion leads us to see time as an unpredictable and innovative factor—a vision much closer to everyday experience than Newton's or Einstein's vision. A simple mathematical model, the baker's transformation, has helped us understand how this notion of time can arise in a purely deterministic world. In celestial mechanics we have observed phenomena more akin to die throwing than to the predictability of Keplerian orbits. In this kind of situation the challenge to the scientist is very much like that of giving an accurate picture of a stream, with its ever-changing flow, its currents and eddies.

This is now an open universe, where time roams freely. Let us quote Heraclitus again: "No one can step twice into the same river." This universal stream carries some recognizable shapes, which one can try to salvage and bring to the shore. This is what our memory does, salvaging a few disjointed recollections from the continuous flow of time and storing them in some unconscious part of our minds. This is also what catastrophe theory is trying to do. Cusps, umbilics, or butterflies arise from special dynamical systems, and the expert will recognize them as they flow past him, even if he cannot explain their genesis or predict their occurrence.

This amounts to a return to the geometrical standpoint we started from. For the elementary catastrophes are nothing but geometrical figures, which hide some very rudimentary dynamics. Note, however, that geometry plays a much less ambitious role in catastrophe theory than it did in Newton's or Einstein's cosmology. It is no longer supposed to provide a complete and detailed model for the physical universe. At best, it should provide a frame of reference, making it easier to grasp certain situations when they occur. It is a very meager role indeed, just one step short of leaving the stage. What a fall for an actor who has played the leading role for so long!

In fact, mathematics keeps oscillating between two very different notions of time. The first is a global notion, most conveniently expressed in geometrical language. The present carries both the past and the future, however distant, like faraway galaxies which make their presence felt by the gravitational pull they exert on molecules on earth. From the second standpoint, time is but a succession of fleeting states, largely independent of each other. The traces of the past disappear very quickly, and each instant of time brings something new.

The true nature of time evades mathematicians. All they can do is to express in their language the tension between these two extremes. They are not alone in doing that; poets and philosophers have also encountered the dilemma, and expressed it in their own, sometimes far better, way. Shakespeare immediately comes to mind:

> Put out the light, and then put out the light:
> If I quench thee, thou flaming minister,
> I can again thy former light restore,
> Should I repent me; but once put out thy light,
> Thou cunning'st pattern of excelling nature,
> I know not where is that Promethean heat
> That can thy light relume.

There is no better way to express the irreversibility of time. Shakespeare rightly points out that it is due to the complexity of natural systems (Desdemona's body) as opposed to the simplicity of manmade ones (a candle). This is precisely what makes celestial mechanics intractable, whereas Kepler's laws, which deal with two-body systems only, are so simple.

We could also quote Macbeth's famous monologue "Tomorrow and tomorrow and tomorrow . . .," or Cassius's words:

> The fault, dear Brutus, lies not in our stars,
> But in ourselves, that we are underlings.

What better way to assert that determinism leaves plenty of room for individual freedom, and that a small-scale event (a knife in Caesar's body) may have large-scale implications (political freedom for millions). English-speaking readers will be much more familiar with all this material than I, so I prefer to leave Shakespeare for Homer.

This single author is responsible for two masterpieces, the *Iliad* and the *Odyssey*. The *Iliad* is a story about youth and going away; the *Odyssey* is a story of middle age and homecoming. Each of them takes up one of the two standpoints from which we see time, so that they serve to illustrate the basic opposition which is the theme of our book.

In the *Odyssey*, time is a single block, from beginning to end. The present announces the future and accomplishes the past. Ulysses looks several times into the future, and bears the consequences of past mistakes.

The whole poem looks to Ulysses' return, νόστιμον ἦμαρ, the day of homecoming, which is also the day of reckoning and in many ways the Last Judgment. Every single day since Ulysses left Ithaca, men and gods have hoped that today would be the day—and this wait has lasted twenty years when the poem begins. Expectations are so high that the day seems close at hand—tantalizingly so, for it never actually happens.

In the first part of the poem, Ulysses never appears; all we hear about him is the question, "When will he return?" His son Telemachus travels to his father's friends, long since back from Troy, Nestor in Pylos, Menelaus in Sparta, to ask that question. In Ithaca the time is ripe; the stage is set for Ulysses' return. When we actually meet the hero, he is at the Phaeacians', a people of seatravelers, asking to be brought home. In the process, he tells of his adventures, the fall of Troy, the Cyclops, and so many others, to a spellbound audience as night falls on Alcinous's palace. At the crucial moment, as he relates his descent to Hades, the kingdom of the dead, when he is about to meet murdered Agamemnon, he breaks his narrative to ask once more for homecoming: "My departure is in your hands, and in the gods' " (11.332).

The day of homecoming, when Ulysses finally sets foot in Ithaca again, is central to the poem in many ways. Strangely enough, Ulysses will never see it, and all witnesses to this event will disappear. Ulysses is asleep when his ship reaches Ithaca.

the future but just concludes the instant, like a door closing:
"hus did the Trojans bury Hector, tamer of horses" (24.804).
The plot is quite simple. It relies on two personal decisions
Achilles. The first is to withdraw to his tent after the quarrel
h Agamemnon. He could have killed Agamemnon: on Ath-
's advice he restrains himself, showering Agamemnon with
ults. This first decision may not be wholly his own, since
ena's intervention was pretty strong (she pulled him back
he hair). But the second, and most crucial one, certainly
ne decides to avenge Patroclus by killing Hector, although
knows that his own death must follow shortly. He could
spared Hector, while avenging Patroclus on the other Tro-
, and sailed back home for a long and prosperous life. This
aat his mother urges him to do, but he chooses the narrower
It is quite an unexpected decision, and Achilles will stick
until the very last moment, when he holds under his spear
or, imploring for mercy.
chilles' third, and last, decision is even more unexpected.
urrenders Hector's body to his father, the Trojan king
n, who has come to his tent at night to beg for it. This
on must be considered in context. Achilles was still under
ock of his friend's death. Not only had he killed Hector,
aany more Trojans besides; he had also spent three days
ing Hector's body by the heels under the ramparts of
nd proclaimed his intent to deprive it of a decent burial.
beautiful scene, when the old king and the young warrior
ogether, and Achilles again reaches a decision which has
ndation in his past experience or in any hopes for the

is the second way to understand time: the present does
lly depend on the past or on the future, and each elapsed
creates something new. The Achilles of the *Iliad*, who
ly in the present, is pitted against the Ulysses of the
who seeks counsel in the past and tries to outguess
are. This is very much like the opposition between
indeterminism and classical determinism, which I have
depict in this book. On one side we have a continual
ng, where the present builds history in an unpredict-

The Phaeacians carry him to the beach and leave. Poseidon will turn the ship and its crew to stone as it enters the Phaeacians' harbor. As for Ulysses, he wakes up the next day and finds himself on a foreign beach, surrounded by the Phaeacian's presents: he does not even recognize his homeland. He hides his possessions, disguises himself as a beggar, and walks inland. No trace is left of Ulysses' homecoming; no one is even aware that the long-sought-for moment has finally happened. What a beautiful symbol: the precise moment when things happen, the instant when the future becomes the past, always slips through our fingers.

So, even when Ulysses is back in Ithaca, and his return has already happened, it will continue being announced as if it were still in the future. There are signs and prophecies until the very end: the seer Theoclymenus predicts the pretenders' slaughter moments before it actually begins.

Ulysses' return is similar to the veil Penelope was weaving in daytime and undoing at night, to put off the pretenders: it is never really over. The connection between the two is even closer: the veil is completed at the same time Ulysses returns. It is because she has had to finish her veil that Penelope resorts

FIG. 4.1. Penelope at her web, conversing with one of the suitors

to the trial of the bow, thereby giving Ulysses the means to finish off the pretenders and complete his homecoming.

The *Odyssey* looks to Ulysses' homecoming, and this day is always thrown farther away into the future: so the whole poem keeps looking toward the future. But it also looks toward the past. It has been ten years since the fall of Troy, but survivors and ghosts alike advise the living and influence the course of events. Nestor and Menelaus give Telemachus whatever news and advice they have, and send him back to Ithaca. Ulysses himself always draws from his Trojan experience. Agamemnon, whose own homecoming in Argos ended in murder, teaches Ulysses to be wary, even of his wife. Ulysses descends into Hades to be counseled by the dead seer Tiresias. And over and over again, be it at Ithaca or in Alcinous's palace, the aeds sing the praise of fallen heroes in battles long past, thereby keeping the memories of the Trojan war alive and rooting current events in that tradition.

There are many occasions in the *Odyssey* when past and future mingle so intimately that it is hard to tell them apart. Remember Tiresias, for instance, the seer long since dead: he is the one whose memories go farthest back into the past, long before the actors of the Trojan war were born, but he is also the one who looks farthest into the future, when the pretenders will be dead and Ulysses will have to walk inland, carrying an oar upon his shoulder, to seek atonement. Remember Penelope, who has managed to remain faithful all these years: she is completely immersed in the past, and yet her actions prove decisive in bringing the homecoming to its conclusion. She retrieves from an attic, where she keeps memories of past glories, Ulysses' great bow, and this relic from the past becomes the means by which Ulysses finally secures his return home.

Past and future are mirror images of each other. Between them, the present disappears, like the day of homecoming, which comes upon Ulysses during his sleep. In the *Odyssey* time is fully predictable: it brings events to their foregone conclusion. As the poem begins, Athena is already predicting Ulysses' return and the pretenders' death. This prediction will be confirmed from many sides as each event brings the epic

closer to its inescapable conclusion. The f
pening, and the past is happening again
of having Ulysses tell his adventures to Al
cians, instead of relating them when t
intricate construction of the *Odyssey*, e
whole as the future mirrors the past. Th
his image in the epic: the aeds, Phemion
hold audiences spellbound and make
himself.

The *Iliad* is quite different. It is an
present is no longer commandeered b
the future with full freedom: "Sing
Peleus' son Achilleus and its devastatio
sandfold upon the Achaians, hurled
house of Hades strong souls of hero
to the delicate feasting of dogs and all
by Richard Lattimore).

These are the opening verses of
acterize it very well. The *Iliad* rela
all fits are. The whole action spans
—a mere instant. The heroes are c
present. Achilles thinks only of his
and then only of avenging him by k
not weigh on the heroes; neither
reward for them. The dead are dea
to fulfill.

And yet both the past and th
tention. The siege of Troy has
and it is high time that it reached
other. Little does Achilles care.
the Greek army is in, and he w
of reconciliation. Only his imme
On the other hand, Achilles
Hector. Once he has commit
Achilles knows he will die, and
Troy. His actions seek no just
the future. The *Iliad* is an isola
its own justification. Its last v

FIG. 4.2. Priam goes to Achilles and offers him a ransom for Hector's body.

able way. On the other, we have an eternal being, where the flow of time is just a human illusion. Gods, or scientists, can see the past and the future at once, as an eternal present, and satisfy themselves that there can be nothing new under the sun.

In between these two notions, we find Thom's ideas: to recognize and salvage some typical shapes from the flow of time. It has its counterpart in literature, too, in the work of Proust. In his search for time past, Proust retains only a few episodes, which, every time they are in tune with the instant his life traverses, will give him an inexpressible joy, and a partial victory in the fight against death:

> The being which had been reborn in me when with a sudden shudder of happiness I had heard the noise that was common to the spoon touching the plate and the hammer striking the wheel, or had felt, beneath my feet, the unevenness that was common to the paving-stones of the Guermantes courtyard and to those of the baptistery of St Mark's, this being is nourished only by the substance of things, in these alone does it find its sustenance and delight. In the observation of the present, where the senses cannot feed it with this food, it languishes, as it does in the consideration of a past made arid by the intellect or in the anticipation of a future which the will constructs with fragments of the present and the past,

fragments whose reality it still further reduces by preserving of them only what is suitable for the utilitarian, narrowly human purpose for which it intends them. But let a noise or a scent, once heard or once smelt, be heard or smelt again in the present and at the same time in the past, real without being actual, ideal without being abstract, and immediately the permanent and habitually concealed essence of things is liberated and our true self which seemed—has perhaps for long years seemed—to be dead but was not altogether dead, is awakened and reanimated as it receives the celestial nourishment that is brought to it. A minute freed from the order of time has re-created in us, to feel it, the man freed from the order of time. And one can understand that this man should have confidence in his joy, even if the simple taste of a madeleine does not seem logically to contain within it the reasons for this joy, one can understand that the word "death" should have no meaning for him; situated outside time, why should he fear the future?[1]

Catastrophe theory is a similar attempt, which is carried on in the field of scientific creation instead of individual psychology. Thom tries to store in the collective unconscious shapes and forms which will serve to recognize classical results and new situations alike, thereby drawing unexpected (and therefore exhilarating) connections across the whole field of knowledge. This may be a hopeless endeavor, and the seven elementary catastrophes probably are much too limited a repertory. But it is well worth trying, and it provides us with a new outlook on time, halfway between Kepler's and Newton's sovereign geometry and Poincaré's constantly flowing chaos. Catastrophe theory salvages a few pieces of floating debris from the shipwreck of geometry.

This is what mathematics tells us about time. But there is still much to be learned from other fields of science. Evolution theory, for instance, studies dynamical systems which have no

1. *Remembrance of Things Past,* trans. C. Scott Moncrieff, T. Kilmartin, and A. Mayor (New York, Random House, 1981), 3:905.

counterpart in physics. They are not deterministic. They would be if the present state of a species completely determined the next state it would reach on the evolution ladder. But such is not the case, for with each generation the species explores the full range of possible evolution, by sending out individuals with different genetic background and by letting the struggle for life select the fittest. This results in such a perfect adaptation to the environment that it is usually mistaken for goal-oriented evolution. Since the eye is so accurately adapted to the function of seeing, it is easy to believe that it was somehow designed for that purpose. The evolutionary process that develops this sophisticated organ from the rudimentary photosensitivity of some individual cells is then understood as implementing a design laid down beforehand. This is an illusion, due to lack of perspective, which would consider the stage reached with the appearance of man as the perfect one, and which would make evolution stop hereafter. It should be rejected, because it is not needed. A much better understanding is gained by considering the interplay between the environment and the genetic stock carried by the individuals. The eye is a product of evolution, not a goal.

This is a point which Stephen Jay Gould, for instance, takes up over and over again. We borrow a striking example from him. In Ireland there are fossil remains of an elk species the male of which carried oversized antlers, spanning seven feet or more. These huge antlers clearly were more of a hindrance than a help to the survival of the species, and from the finalistic point of view, if it is assumed that evolution aims for a better adaptation to the environment, it is hard to understand why they were developed at all. On the other hand, if one tries to understand how the genetic stock reacts to the environment, without any particular purpose in mind, everything becomes much clearer. The dimension of the antlers is genetically linked with other individual characters, such as the overall size of the animal. They cannot vary independently; any change in one of them entails a change in the others. So any modification of these particular genes will have some positive and some negative effects, the question then being whether the positive ones will

balance the negative ones. Apparently, increasing the size of
the animal carried more advantages than oversize of the antlers
held drawbacks. In addition, the dimension of the antlers is a
secondary sexual characteristic, the effect of which is to attract
females and frighten away rivals. The individual with the larg-
est antlers thus has better access to females, and stands a better
chance to reproduce his genes. So the internal logic of evolution
encouraged the overdevelopment of antlers, even though a species with smaller antlers, or no antlers at all (females do very
well without them), would probably have been better suited
for survival.

So evolution continues in the same direction, until a stage
is reached when the survival of the species is jeopardized—the
environment changes from open grasslands to dense forests,
or the antlers reach such a size that the animal can hardly lift
its head. Other directions must then be explored; if none is
found, the species dies out. If a new direction appears, it will
be followed until evolution again reaches a stage where the
species is endangered, and the process of trial and error begins
again.

Besides deterministic and stochastic systems, there seems
to be room for at least one other type of dynamical system,
which is closer to evolutionary processes and may therefore be
called Darwinian. In such systems the past history determines
various possibilities for the present state, among which the
system will choose in response to external stimuli, such as
changes in the environment. Each evolutionary stage looks like
the final goal toward which the species was striving; but this
illusion is shattered as the next stage sets in, and the "final"
state is shown as one in an endless succession of states, a step
in an aimless march to infinity.

These are just a few of the many attempts to portray time.
This evasive model leaves only a facet of its rich personality at
each sitting, and the truth must lie in the confrontation of all
these portraits. Do we not know, from our everyday experience,
that time is constraining but leaves room for freedom, that it
is subject to compelling regularities, although each instant brings
us something new? Proust has taught us to live the past and

the present together, as a simple experience, when an everyday occurrence unexpectedly calls forth from the depths of the unconscious some forgotten memories. This is one way for the individual to conquer death, to turn back, even for one moment, the flow of time, which ultimately will carry us to our end. There is no deeper problem for the individual. Our attempts to master time are aspects of our fight against death, and it is no wonder that we find in science similar attempts, reflecting the same anxiety.

In Madrid, at the Prado museum, there is a small panel of Hieronymus Bosch which represents the temptation of Saint Anthony. It is bathed in an otherworldly light, diffuse but crystal clear, which leaves no shadows but creates reflections. The light seems to stream from the background, where strangely modern skyscrapers rise in the distance. An old hermit sits in the foreground, by a stream, his back to a hollow tree. His brown robe and the green bank stand out vividly against the light colors, mostly ochers, which stretch out behind him. A strangely peaceful landscape occupies the middle ground; a gate in a hedge opens onto a path which runs into the woods; a canal cuts across the center of the picture, crossed by a bridge which leads to a rustic chapel and a clump of trees.

Three trees rise on the bank. Their trunks, straight and smooth, enclose Saint Anthony in the foreground, cutting him off from this beautiful landscape. His back is turned to it, his folded hands and his head reclining on his staff, in a meditative posture. He gazes at the flowing stream, which carries strange creatures, partly animal and partly mechanical. Some of these monsters have climbed the bank and advance toward him. Others have landed upstream and attack the luminous landscape in the background.

We are in a similar position. We turn our backs to the world we belong to, like the prisoners in Plato's cave, watching the shadows on the walls. We cannot gain a direct understanding of reality, any more than the prisoners can break their chains or the hermit stand up and turn to the landscape that stretches out behind him. Our gaze is turned to the stream of time—or rather to that small stretch of its flow by which we sit and

FIG. 4.3. Hieronymus Bosch, the temptation of Saint Anthony. Photo, Alinari/Art Resource, N.Y.

which we call the present. This contemplation is science itself. It creates strange monsters, which invade the whole world and eventually turn against us.

But the hermit is lost in contemplation. His mind has touched the land of eternal truth. His assailants feel that he is out of reach, and their efforts become perfunctory. He does not see the two-legged tower which squats at his side, wielding a mallet, or the menacing creatures which approach behind his tree. The arrow aimed at him will miss, and the devil's claw will not reach him. All above him, the sky spans blue.

APPENDIX 1
PRELUDE AND FUGUE
ON A THEME BY
POINCARÉ

The theme which we shall develop is shown in Figure A1. Poincaré was the first to draw this picture. He discovered it in the course of the investigations which we related earlier. As we saw in chapter 2, one of his ideas was to replace dynamical systems in space by point transformations in the plane. If a closed trajectory of the dynamical system is known, the situation in its immediate neighborhood can be investigated by positioning a plane across it, cutting it at a point O. Neighboring (not necessarily closed) trajectories will then be represented by their intersection points with some transverse plane, which constitute a doubly infinite sequence (toward the past and toward the future). This situation is depicted in chapter 2 in Figure 2.3.

Figure A1 shows one of these doubly infinite sequences, as it appears in the transverse plane. The first intersection point is M_0, obtained at time $t = 0$, the next one is M_1, then M_2, M_3, and so on, as time increases. If we move backward in time, we meet successively M_{-1}, M_{-2}, M_{-3}, and so on. The points M_n are called *iterates* of the initial point M_0. The first ones are the positive iterates, the second ones the negative iterates. They are all located on the same trajectory of the dynamical system we are investigating. By losing one dimension, we have replaced a curve in three-dimensional space by a sequence of points in the plane.

The point O lies at the intersection of this plane with the closed trajectory we are starting from. Since the corresponding solution is periodic, all subsequent intersection points will also lie at O. In other words, the points O_1, O_2, . . . , and O_{-1}, O_{-2}, . . . , all coincide with O.

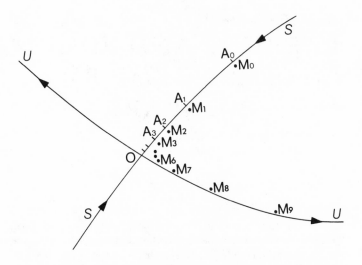

FIG. A1. The stable curve S and the unstable curve U intersect at
O, the fixed point.

We define a point A_0 to be stable if its *positive* iterates converge to
O, that is, if the successive intersection points A_n get ever closer to
O as the integer n takes higher and higher positive values. The set S
of all stable points is depicted in Figure A1: it is seen to be a curve
going through O, which we shall refer to as the stable curve. If a
point A_0 lies on S, so do all its iterates A_n, and the positive iterates
approach O on S as n increases to $+\infty$, while the first few negative
iterates move away from O and A_0 on S.

Similarly, a point B_0 will be unstable if its *negative* iterates B_n
converge to O, that is, if the sequence $B_{-1}, B_{-2}, B_{-3}, \ldots$, approaches
O as the index takes larger and larger negative values. So B_0 is the
present position of a point which was very close to O a long time ago.
The set of all unstable points is a curve U going through the origin,
which we will refer to as the unstable curve. If B_0 lies on U, so do
all its iterates. Its first few positive iterates $B_1, B_2, \ldots$, drift quickly
away from O and B_0; hence the reference to instability.

To summarize the previous discussion, a point will be stable if it
reaches O in infinite time, and unstable if it drifts away from O in
infinite time. If we pick a point in the plane which is neither stable
nor unstable, say M_0, close to S but not on S, we see that at first

its positive iterates M_n drift slowly away from S, while following the general trend toward O which characterizes stable points. But as soon as M_n gets close to O, the influence of the unstable curve U begins to be felt, and the next iterates run away from O in the direction of U.

The stage is set, the theme has been played; let us now start the show. We have to develop the suggested theme, that is, to unravel all its hidden complexities. Is Figure A1 as simple as it seems? For instance, can a point be both stable and unstable? The only way to know is to extend U and S beyond the loose ends in Figure A1 and see what happens.

There are several possibilities, depicted in Figures A2–A5. The first two illustrate very particular situations, where the unstable curve emanating from O becomes the stable curve for another point O' (Fig. A2), or for the point O itself (Fig. A3). In Figure A4, the curves U and S run away to infinity. In many cases, physical considerations, such as conservation of energy, will confine all iterates of a point within some finite region, so that U and S will be confined also, and the situation in Figure A4 cannot happen.

We are left with the last, most interesting case, the one Poincaré analyzed: the curves S and U intersect transversely, as in Figure A5. The intersection point, H, is called a *homoclinic point*.

This innocent-looking picture is going to explode under our eyes. The point H belongs to the stable curve S. Its positive iterates H_1, H_2, . . . also belong to S, and they converge to O.

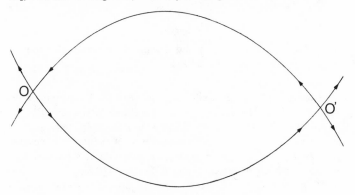

FIG. A2. There are two fixed points, and the unstable curve for one is the stable curve for the other.

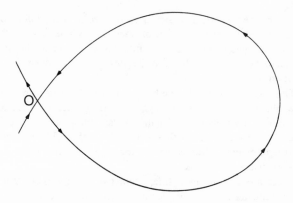

FIG. A3. The stable and unstable branches out of *O* merge into the same curve.

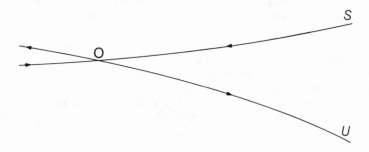

FIG. A4. The stable curve S and the unstable curve U run away to infinity without intersecting again.

The point *H* belongs to the unstable curve U. All its iterates also belong to U. So $H_1, H_2, \ldots$ must belong both to U and to S.

In other words, the curves U and S must intersect, not only at *H* but at $H_1, H_2, \ldots$, an infinite sequence of points. The negative iterates, $H_{-1}, H_{-2}, \ldots$, constitute another infinite sequence of intersection points, which converges to *O* along U.

We started with a single homoclinic point, and lo and behold, we now have infinitely many. The fun does not stop here, because all these new homoclinic points have to be located on the curves S and U to which they belong. This means that we have to extend U beyond

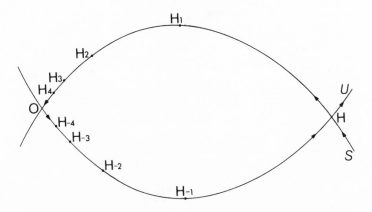

FIG. A5. The stable and unstable curve out of O meet at a point H, called a homoclinic point. The picture shows its first forward and backward iterates.

the loose end hanging in Figure A5, and show how it intersects S at H_1, H_2, and all the positive iterates.

Proceed naively, as in Figure A6. This has to be wrong. For if we pick on U a point P close to H, its positive iterates $P_1, P_2, \ldots$ will remain on U. On the other hand, P is close to a point of S, namely, H. We know, then, how the positive iterates should behave: they will stay close to S until they get sufficiently close to O, and then they will run away along U. This means that eventually some iterate P_n, with large positive n, will cross S along U, to the right of the picture, as in Figure A7.

But P_n belongs to an arc $H_n H_{n+1}$ whose extremities H_n and H_{n+1} are located on S, in the immediate neighborhood of O. This arc will have to stretch tremendously along U to reach P_n. It must cross S near H in two points M and Q. Of course the next arc $H_{n+1}H_{n+2}$ will be even more stretched, the next one $H_{n+2}H_{n+3}$ still more, and the complexities of the situation are already beyond our capacity to draw.

It is not over yet. The points M and Q are new homoclinic points which are not iterates of H. They do not belong to the doubly infinite sequence H_n which we have already discovered.

By the same arguments as before, we see that the positive $(M_1, Q_1), (M_2, Q_2), \ldots$ and negative $(M_{-1}, Q_{-1}), (M_{-2}, Q_{-2}), \ldots$ iterates will again be homoclinic points. This means that not only the arc $H_n H_{n+1}$ but all its positive and negative iterates, that is, all arcs in

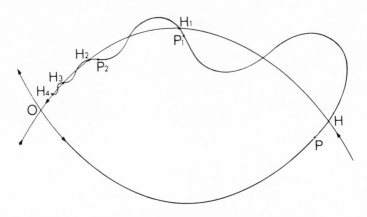

FIG. A6. First attempt at continuing U beyond *H*.

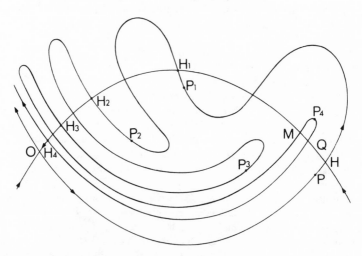

FIG. A7. Second attempt at continuing U.

the doubly infinite sequence . . . , $H_{-2}H_{-1}$, $H_{-1}H$, HH_1, H_1H_2, . . .
must intersect S in two points.

For arcs that succeed H_nH_{n+1} on U, it is clear enough how this is
going to happen: they will pile up on the arc *OH* of U, and the
intersection points will converge to *H* on the arc *OH* of S. But arcs

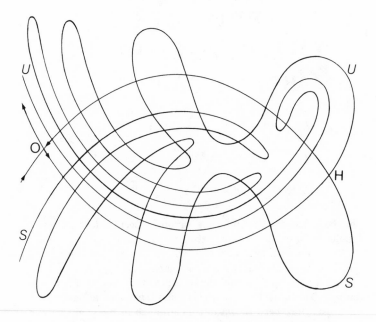

FIG. A8. Third and last attempt at continuing U. The drawing is incomplete: the curve U must fold and accumulate indefinitely along the arc OH of S, so that a host of new homoclinic points appears.

that come earlier, $H_{n-1}H_n$ or H_0H_1, for instance, cannot meet S between O and H. So S has to meet them at some point later than H, that is, the curve S has to fold back along itself to intersect all the earlier arcs.

From the modest beginning of Figure A1, the stable and unstable curves have knitted a web, intricate beyond our wildest imagination. As Poincaré put it: "The reader will be struck by the complexity of this figure, which I am not even attempting to draw. Nothing could give us a better idea of the intricacy of the three-body problem, and of most problems in dynamics."[1]

1. *Méthodes nouvelles*, vol. 3, chap. 3, sec. 397.

APPENDIX 2
THE FEIGENBAUM
BIFURCATION

This book has tried to introduce various concepts, such as chaos, periodicity, or equilibrium, in a geometrical way. These concepts can also be approached in a computational way. All one needs is a small hand calculator, preferably a programmable one.

In a series of papers published in 1978, the physicist M. J. Feigenbaum has called attention to a very simple one-dimensional model, which has since been shown to contain much of the complexity of dynamical systems. It is the transformation of the interval $[-1, 1]$ into itself, which associates with every point x the point $1 - \mu x^2$. This transformation depends, of course, on the value of the parameter μ, which will be chosen between 0 and 2.

Once μ is chosen, we can iterate this transformation. This means that we pick an initial point x_0, the transform of which is $x_1 = 1 - \mu x_0^2$, and then proceed with transforming x_1 into x_2, then x_2 into $x_3, \ldots, x_n$ into x_{n+1} according to the rule

$$x_{n+1} = 1 - \mu x_n^2.$$

Let us try it.

1. Values of μ between 0 and 0.75.

Let us take $\mu = 0.5$, for instance. The reader is invited to try any other value of μ in the same range.

Here is a first series of results, starting from $x_0 = 0$:

$$x_0 = 0$$
$$x_1 = 1$$
$$x_2 = 0.5$$

$$x_3 = 0.875$$
$$x_4 = 0.6171875$$
$$x_5 = 0.809539795$$
$$x_6 = 0.67232266$$
$$x_7 = 0.77399112$$
$$x_8 = 0.700468872$$
$$x_9 = 0.754671679$$
$$x_{10} = 0.715235328$$
$$x_{11} = 0.744219212$$
$$x_{12} = 0.723068881$$
$$x_{13} = 0.738585696$$
$$x_{14} = 0.727245584$$
$$x_{15} = 0.735556929$$

and so on, to:

$$x_{20} = 0.731312469$$
$$x_{25} = 0.732205977$$
$$x_{30} = 0.732018182$$

The iterates converge to a limit point:

$$x_\infty = 0.732050807.$$

We could try another initial point. Take, for example, $y_0 = -0.5$, and see what happens:

$$y_0 = -0.5$$
$$y_1 = 0.875$$
$$y_2 = 0.6171875$$
$$y_3 = 0.809539795$$
$$y_4 = 0.67232266$$
$$y_5 = 0.77399112$$
$$y_{10} = 0.723068881$$
$$y_{15} = 0.733930922$$
$$y_{20} = 0.731655187$$
$$y_{25} = 0.732133965$$
$$y_{30} = 0.732033324$$

The iterates converge to a limit point:

$$y_\infty = 0.732050807.$$

This is the same result as the one above. We are dealing with a dissipative system, as defined in chapter 3: whatever the initial position x_0 may be, the system will eventually come to rest on the point 0.732050807, which is a stable equilibrium.

So, for $\mu = 0.5$, the system has a stable equilibrium at the point 0.732050807. It can be shown that, for all values of the parameter μ between 0 and 0.75, the system has a stable equilibrium x_μ, the position of which depends continuously on μ. In fact, x_μ is given in terms of μ by the explicit formula:

$$x_\mu = [-1 + \sqrt{(4\mu + 1)}]/2\mu.$$

Setting $\mu = 0.5$, we obtain

$$x_\mu = -1 + \sqrt{3} = 0.732050808,$$

which coincides with the value found experimentally for x_∞ and y_∞.

2. Values of μ between 0.75 and 1.25.

Take $\mu = 1$, for instance. Once again, the reader is invited to try another value of μ within the same range.

Here is a first series of values, obtained from $x_0 = 0$:

$$x_0 = 0$$
$$x_1 = 1$$
$$x_2 = 0$$
$$x_3 = 1$$

One quickly finds out that the values are alternatively 0 and 1. That is, we have found a closed trajectory, the period of which is 2.

This may just be a fluke: perhaps it is the only periodic solution, and we just happened to hit upon it. To find out, we should try another starting point, for example, $y_0 = 0.5$:

$$y_0 = 0.5$$
$$y_1 = 0.75$$
$$y_2 = 0.4375$$
$$y_3 = 0.80859375$$
$$y_4 = 0.346176147$$
$$y_5 = 0.880162075$$
$$y_6 = 0.225314721$$
$$y_7 = 0.949233276$$

$$y_8 = 0.098956187$$
$$y_9 = 0.990207673$$
$$y_{10} = 0.019488764$$
$$y_{11} = 0.999620188$$
$$y_{12} = 0.0007594796$$
$$y_{13} = 0.999999423$$
$$y_{14} = 0.0000011536$$
$$y_{15} = 1$$
$$y_{16} = 0$$

This is no longer a closed trajectory, but it converges very quickly to the closed trajectory we have just found. From the fifteenth iteration on, the points x_n and y_n are in fact so close that the computer is unable to tell them apart, and simply registers the value y_n to be 0 if n is even and 1 if n is odd. This is a round-off error, due to the fact that the calculator performed the computations with nine decimal places only. A more precise computation would be able to follow y_n for a longer time, but eventually the same problem would appear.

It should be noted that the point x_μ given by the formula

$$x_\mu = [-1 + \sqrt{(4\mu + 1)}]/2\mu$$

still is an equilibrium. For $\mu = 1$ we get $x_\mu = 0.618033988$. If we enter $x_0 = 0.618033988$ into the computer, we get the same value indefinitely: $x_n = 0.618033988$ for all values of n.

But it is now an unstable equilibrium! To see this, let us move very slightly away from it by changing the last decimal digit from 8 to 9. This is a modification of one part in one billion. Now watch how it drifts away, slowly at first, and then quickly:

$$y_0 = 0.618033989$$
$$y_1 = 0.618033988$$
$$y_2 = 0.618033989$$
$$y_3 = 0.618033988$$
$$y_4 = 0.618033989$$
$$y_5 = 0.618033988$$
$$y_6 = 0.618033989$$
$$y_7 = 0.618033987$$
$$y_8 = 0.61803399$$
$$y_9 = 0.618033987$$

$$y_{10} = 0.61803399$$
$$y_{11} = 0.618033986$$
$$y_{12} = 0.618033991$$
$$y_{13} = 0.618033985$$
$$y_{14} = 0.618033993$$
$$y_{15} = 0.618033983$$
$$y_{16} = 0.618033995$$
$$y_{17} = 0.61803398$$
$$y_{18} = 0.618033999$$
$$y_{19} = 0.619033975$$
$$y_{20} = 0.618034005$$
$$y_{21} = 0.618033968$$
$$y_{22} = 0.619034014$$
$$y_{23} = 0.618033957$$

Note that odd-numbered terms are smaller than x_μ and decrease, while even-numbered terms are greater than x_μ and increase. The first converge to 0, the second to 1, as the rest of the computation shows:

$$y_{99} = 0.065162952$$
$$y_{100} = 0.995753789$$

The first six terms of the sequence just oscillate between 0.618033989 and 0.618033988. This is because the calculator only displays nine decimal places, even though it performs the computations with more. The display is just the tip of the iceberg, and it is the immersed part, the undisplayed tail of the number, which sways more and more during these first few steps and destabilizes the system.

3. Values of μ between 1.25 and 1.368.

Take, for example, $\mu = 1.3$. The reader will discover for himself a closed trajectory of period 4, toward which all other trajectories converge. It goes in succession through the points:

$$-0.01494637$$
$$0.999709587$$
$$-0.29924503$$
$$0.88358813$$

The point 0.573069199 is an unstable equilibrium. There is also an unstable closed trajectory of period 2, going through the points

$$[1 + \sqrt{(4\mu - 3)}]/2\mu = 0.955092191,$$
$$[1 - \sqrt{(4\mu - 3)}]/2\mu = -0.18586142.$$

Let us now pause and look back. We have crossed two catastrophes. For we are in the general situation of chapter 3, namely, a dynamical system depending on a parameter μ. As long as the parameter remains in the range [0, 0.75], the qualitative behavior of the system is unchanged: there is a stable equilibrium x_μ, depending continuously on μ, toward which all trajectories converge. Similarly, when the parameter remains in the range [0.75, 1.25], the qualitative behavior of the system is unchanged: there is a stable closed trajectory of period 2, toward which all other trajectories converge. But crossing the value $\mu = 0.75$ changes the qualitative behavior: the stable equilibrium disappears (or, rather, it loses its stability and all dynamical significance), and a stable 2-periodic trajectory appears instead. So $\mu = 0.75$ is a catastrophic value, according to the general definition of chapter 3 (but not in the restricted sense of elementary catastrophe theory, since the system under consideration ceases to be dissipative).

Similarly, $\mu = 1.25$ is a catastrophic value, since the 2-periodic trajectory loses its stability in favor of a 4-periodic trajectory.

4. Values of μ between 1.368 and 1.401.

We find a period-doubling cascade. That is, there is an infinite sequence of catastrophic values μ_n increasing and converging to 1.401:

$$1.368 = \mu_2 < \mu_3 < \ldots < \mu_n < \mu_{n+1} < \ldots < 1.401,$$

such that, if μ lies between μ_n and μ_{n+1}, the system has a closed trajectory of period 2^{n+1} toward which all other trajectories converge. In other words, crossing these catastrophic values in the direction of increasing μ means that the period of the stable closed trajectory is doubled.

The μ_n can be located by the following relation, which determines their rate of convergence:

$$1.401 - \mu_n = 4.669 \, (1.401 - \mu_{n+1}).$$

The number 4.6992 . . . is the Feigenbaum constant. Its appearance in many different circumstances has been one of the great scientific puzzles of past years. It seems to have some deep physical significance for bifurcation phenomena.

5. Values of μ between 1.401 and 2.

This region is less well understood. The dominant facts are as follows:

a) For most values of μ in that region, the system has a chaotic behavior. All the periodic trajectories one can find are unstable, and the system just wanders aimlessly from one end of the interval [−1, 1] to the other. The reader is invited to pick his own value for μ and x_0, and to have a look at the iterates. He will most likely observe a random succession of values.

b) In this vast chaotic sea, there are a few islands of order and stability. The reader should check the region [1.75, 1.7685]. We will not spoil the surprise by telling him what he will find.

This close mingling of order and chaos, this progressive transition from one to the other by period-doubling, these islands of periodicity in a chaotic sea, all this should call to mind Figures 2.4 and 2.5 of chapter 2, and Poincaré's beautiful arguments. Order and chaos are aways found together, be it in celestial mechanics or in number crunching.

BIBLIOGRAPHY

This is an unpretentious and incomplete bibliography, the only aim of which is to acknowledge a few books from which the author has derived some inspiration.

CHAPTER 1

Koyré, Alexandre. *La Révolution astronomique*. Paris: Hermann, 1961.
————. *Newtonian Studies*. Cambridge: Harvard University Press, 1965.
————. *Etudes Newtoniennes*. Paris: Gallimard, 1968.
————. *Astronomical Revolution*. Ithaca, NY: Cornell University Press, 1970.

CHAPTER 2

Poincaré, H. *Les Méthodes nouvelles de la mécanique céleste*, vols. 1–3. Paris: Gauthiers-Villars, 1892–99.
Prigogine, I., and Stengers, I. *La Nouvelle Alliance*. Paris: Gallimard, 1979.

CHAPTER 3

Thom, René. *Stabilité structurelle et morphogénèse: Essai d'une théorie générale des modèles*. Reading, MA: Benjamin, 1972. (English version: *Structural Stability and Morphogenesis*, tr. D. H. Fowler. Reading, MA: Benjamin, 1974.)
————. *Paraboles et catastrophes*. Paris: Flammarion, 1983.

CHAPTER 4

Delevoy, R. *Bosch*. Lausanne: Skira, 1960.
Finley, J. *Homer's "Odyssey."* Cambridge: Harvard University Press, 1978.

INDEX